Compressors

About the Authors

Heinz P. Bloch is a Registered Professional Engineer. In his 50-year machine design and reliability improvement career he completed assignments as Exxon Chemical's Regional Machinery Specialist for the United States. Mr. Bloch has also held machinery-oriented staff and line positions with Exxon affiliates in the United States, Italy, Spain, England, The Netherlands, and Japan. He has conducted more than 500 public and in-plant courses in the United States, Africa, Asia, Australia, Europe, and South America. Mr. Bloch is the Equipment and Reliability Editor for *Hydrocarbon Processing magazine*.

Fred K. Geitner is a registered Professional Engineer engaged in process machinery consulting. Before retiring from Esso/Exxon Mobil after more than three decades, he held long-term assignments as Exxon Chemical's Regional Machinery Specialist for Canada and machinery-oriented staff and line positions with Exxon affiliates in France. Reliability improvement missions have taken Mr. Geitner to process plants and manufacturing facilities in at least 17 countries, and he continues to conduct public and in-plant courses worldwide.

Compressors
How to Achieve High
Reliability & Availability

Heinz P. Bloch

Fred K. Geitner

New York Chicago San Francisco
Lisbon London Madrid Mexico City
Milan New Delhi San Juan
Seoul Singapore Sydney Toronto

The McGraw·Hill Companies

Cataloging-in-Publication Data is on file with the Library of Congress.

McGraw-Hill books are available at special quantity discounts to use as premiums and sales promotions, or for use in corporate training programs. To contact a representative please e-mail us at bulksales@mcgraw-hill.com.

Compressors: How to Achieve High Reliability & Availability

1 2 3 4 5 6 7 8 9 0   DOC/DOC   1 9 8 7 6 5 4 3 2

ISBN 978-0-07-177287-7
MHID 0-07-177287-1

The pages within this book were printed on acid-free paper.

Sponsoring Editor	**Copy Editor**	**Composition**
Michael Penn	Medha Joshi	Cenveo Publisher Services
Acquisitions Coordinator	**Proofreader**	
Bridget Thoreson	Barnali Ojha	**Art Director, Cover**
		Jeff Weeks
Editorial Supervisor	**Indexer**	
David E. Fogarty	Robert Swanson	
Project Manager	**Production Supervisor**	
Sandhya Gola, Cenveo Publisher Services	Richard C. Ruzycka	

This book is an expression of respect for those who diligently design, operate, maintain, and often manage large industrial facilities. Some of these value-adding men and women do so against considerable odds, and many persist with admirable skill, fairness, constancy, and great technical competence. We honor them.

Contents

Preface

After we retired from our respective jobs as regional machinery specialists for the petrochemical segment of one of the world's largest multinational petrochemical and oil refining corporations, we teamed up to write several successful texts on equipment reliability and failure analysis. Our writing tasks were facilitated by a similar career focus on machinery failure avoidance. Fred Geitner represented the corporate reliability interests in Canada and Heinz Bloch did similar work in the United States. As coauthors we carried over into "semi-retirement" what we had learned in over 100 man-years of work exposure. We did indeed spend 100 man-years in industry after graduating from technical universities as mechanical engineers in the early 1960s. After authoring books on process machinery failure analysis and troubleshooting in the 1980s and 1990s, we came up with the idea of doing a book on "Compressors: How to achieve high reliability & availability." The emphasis is clear: *How to achieve reliability and availability.*

Why another book, and why would someone need *this* text? After all, compressor users have had (and still have) access to hundreds of books, and many thousands of articles dealing with gas compressor subjects. But we also know that an unacceptably large number of air and process compressors fail unnecessarily or even catastrophically every year. Some of the failure causes are elusive, overlooked, undocumented, or hidden in rather voluminous books. As involved compressor specialists, we are under no illusion as to what people do with technical books when they are overwhelmed by their sheer volume. Books are tools, and all tools are useless unless they are being used. Likewise, books are of no value until they are being read. And, to paraphrase Mark Twain, the person who refuses to read is as illiterate as the person who cannot read.

We wanted to explain some compressor issues and clearly map out practical remedies that worked. Remedial or preventive actions must be taken by operators, technicians, engineers, and managers. They are our collective audience and the book was written with an acute sense of audience awareness. In essence, we wanted to leave a legacy of things learned to prevent failures and *yes*, we believe our thousands of pages of previous texts have *not* always managed to do this in sharply focused, well-considered paragraphs. We wanted to keep this text to about 250 pages and had to make a choice of what to cover. So, we decided to limit ourselves to centrifugal and reciprocating compressors. And the 250 pages were picked on the basis of assuming that serious students will find time to read maybe two pages per day. We thought the entire book could then be read in about 125 workdays—perhaps between coffee or tea breaks....As any close review of what has been offered in the past will uncover, many texts were written to primarily benefit one particular job function, ranging from equipment operator to designer. We know that some compressor books are too academic, other are too vague, and some have failed to divulge enough details to be of lasting value. Some books, hopefully none of ours, contain a hidden bias and often appeal to a very narrow spectrum of readers. Others texts are perhaps influenced by a particular user's somewhat unique experience and will simply not apply across a sufficiently wide spectrum.

Again, although we had written other books and articles and conference papers and course manuscripts on compressor reliability improvement, some important material is too widely dispersed to be readily accessible. We therefore set out to assemble, rework, condense, add new items, and explain the most valuable points. We wanted to convey the message in a widely distributed text with, hopefully, permanence and "staying power." To satisfy the scope and intent of this book, we endeavored to keep theoretical explanations to a reasonable minimum, although Fred still knows more math today than Heinz ever knew.... Yet, we collaborated because we wanted to clearly describe applicable issues. These are issues that will continue to encroach on both compressor safety and reliability, unless people out there choose to no longer overlook them.

This text then tries to present a balanced view between compressor design, procurement, and practical compressor use. It certainly intends to do justice to the thermodynamic, aerodynamic, and mechanical issues of interest to all parties. But, again, it will do so without too many theoretical equations. The book briefly lays out how the two principal compressor categories (dynamic and positive displacement) function. The text then quickly moves on to guidelines and details that must be considered by reliability-focused compressor users. A number of risky omissions or shortcuts by designers, manufacturers, and user-operators are also described. They must be processed into specifications that ultimately result in procuring more reliable compressors.

We know this book will help; make good use of it and remember that it reflects the work processes and procedures used by the very best and most profitable companies. That, then, leads us to acknowledge the assistance we received from world-class vendor-manufacturers who graciously agreed to let us include a large number of nonproprietary photos and illustrations: AESSEAL (Rotherham, UK), Cooper-Bessemer (Mt. Vernon, OH), Dresser-Roots (Connersville, IN), Nuovo Pignone (Florence, Italy), Elliott Company (Jeannette, PA), Hitachi Ltd, (Tokyo, Japan), Mitsubishi (Hiroshima, Japan), Borsig (Berlin, Germany), A-C Compression (Appleton, WI), IMO-Demag-DeLaval (Trenton, NJ), Salamone Turbo-Engineering (Houston, TX), Hickham Industries (LaPorte, TX). Some of these companies were the predecessors to what became global entities and we remember them with thanks.

HEINZ P. BLOCH, P.E.
FRED K. GEITNER, P. ENG.

Compressors

CHAPTER 1

Introduction, Compression Principles, and Internal Labyrinths

$\mathbf{M}$ any good compressor texts start out by explaining that these machines depend on thermodynamic laws to operate. Compressor books then typically progress to the study of thermodynamics. Well, this text doesn't do that, because we have not planned to add to the body of literature that delves into the academic and mathematical treatment of compressors. In this book we want to keep strictly in mind the goal of capturing topics that are not readily accessible. We want compressor troubleshooting and repair to deal with failure avoidance topics and issues. We also want to put our spotlight on tricks of the trade that are often overlooked. From personal experience, we know that these items, and especially the spotlighted areas, deserve to be captured. They should be of interest to a very wide spectrum of readers; we call this target audience the *reliability-focused professional community*.

Most of our readers know why compressors are used in modern industry, but we'll briefly recap the obvious reasons for their use. We also wish to cite here a mere handful of general principles for the benefit of readers with little or no prior exposure to compressors. However, we want to be clear as to what this text can do and will do. Access to other learning or "looking up details" may be needed if our comments are too brief.

Density and Compression Ratios

Compressors take a certain amount of gas and increase its pressure from one level (suction pressure) to another level (discharge pressure). In doing so, the volume of the gas will be reduced, its temperature

1

will increase, and its mass will stay the same. A pound of gas at an absolute pressure of 145 psia will have less volume than a pound of the same gas at 14.5 psia, but it's still a pound. Recall that a pound of goose feathers weighs as much as a pound of steel; although the authors admit that they would rather be struck in the head by a pound of goose feathers than a pound of steel. Anyway, mass can neither be created nor can it be destroyed, except in a nuclear reaction, which is really an altogether different branch of physics. Because Albert Einstein has already explained this, we don't have to go there.

Back to our particular example where we have assumed a discharge pressure of 145 psia—pounds per square inch absolute. And, assuming we had a suction pressure of 14.5 psia, the compression ratio would be 145:14.5 = 10:1. Note that we must do these calculations in units of absolute pressure. If we had attempted to do this in units of gage pressure (psig), it would have been 130.5 psig divided by 0 psig. We vaguely remember that it's illegal to divide by zero and we had promised our publishers that we would not do anything illegal. That's why we will always calculate pressure ratios in absolute units.

It is not normally possible to achieve a compression ratio of 10:1 in a single compression stage. If we tried to do that, the temperature of the gas would go far too high. At excessively high temperatures we would have to opt for very expensive metallurgies. Also, too much energy would go into heat and the compression process would be rather inefficient. Moreover, at the resulting high temperatures there would be thermal growth because metals expand when heated and contract when cooled. So, we are limited to stay more or less near an average compression ratio of 3:1 per stage in positive-displacement machines and somewhat lower in centrifugal compressors. With a lighter gas and in centrifugal compressors, the achievable ratios might be closer to 2:1. With heavier gases one might occasionally see compression ratios approaching 4:1.

If we need to achieve a compression ratio of 10:1 but are concerned about high temperatures, we could take the gas discharged from a compression stage and lead it into a heat exchanger. After cooling it we could pipe it to the intake of the next stage and perhaps increase the gas pressure by a ratio of 3.4:1. Now we would perhaps get it to a final discharge pressure of $3 \times 3.4 = 10.2$ times the original suction pressure.

Well, not quite. Actually, we would expect the final discharge pressure to be only 10 times (not 10.2 times) original suction. That's because we figured there was a frictional loss or pressure drop in the heat exchangers, and we assumed that this pressure drop amounted to about 0.2 psi.

Gas density [Eq. (1.1)] varies in close proportion to gas molecular weight:

$$\text{Gas density} = \frac{P(MW)}{R \times T} \qquad (1.1)$$

where P = gas pressure (absolute)
          T = gas temperature (absolute)
          MW = gas molecular weight
          R = gas constant

So, if some gases are denser or thicker than others, it is because they have different molecular weights. Amedeo Avogadro, in 1811, hypothesized that two given samples of an ideal gas, at the same temperature, pressure, and volume, contain the same number of molecules. Thus, the number of molecules or atoms in a specific volume of gas is independent of their size or the molar mass of the gas. While at standard conditions of temperature and pressure, the molecules of two different gases will take up the same volume, the denser gas molecule will be heavier. We might liken this to a ping-pong ball and a golf ball. Both take up the same space, but the golf ball is denser and will weigh more. And the ping-pong ball is easier to compress than the golf ball.

Heat and Mass Concepts Simplified

It is also worth mentioning that some gases build up or retain more heat than others. It's a relationship that is often expressed in a specific heat or k-value. If we really needed to know more about it, we could easily find it explained in books. And common sense always helps. If we applied a finite amount of heat to the lower portions of both a ping-pong ball and a golf ball, we would not expect this heat to travel through each ball at the same rate of speed and the two surfaces might not ever reach the same steady-state temperature. Each dissipates heat differently or radiates it into the surrounding environment at different rates.

Also, the greater the mass flow rate of gas we wanted to elevate from one pressure to a higher pressure, the more energy we would have to put into the gas. Mass flow is throughput; it is often expressed in kilograms per second, which is about the same as 2.2 lb/s. In the context of compressors, energy is generally measured in either kilowatts or horsepower.

The Concept of "Head" and Other Parameters

Another common measure is "compression head," generally abbreviated as "head." *Head* is a term used in computing the amount of energy that must be added to each mass unit of gas to produce the desired pressure increase. The customary units are foot-pounds of energy per pound of gas, which then simplifies to "feet." Head is borrowed from the field of liquid hydraulics where the height of a column of liquid, in feet, is equivalent to the energy theoretically needed to produce a static pressure at the base of the column. The concept is being carried over to compressible gas technology, and pressure ratio is substituted for pressure at the base of the column. Molecular weight is substituted for liquid specific gravity and an arrangement of exponents [the n in Eq. (1.2)] is used. Suitable exponents account for both the thermal behavior of a particular gas and the unavoidable inefficiency of a compressor.

$$H_p = \frac{1545}{MW} Z_{AVG} T_1 \left(\frac{n}{n-1} \right) \left[\left(\frac{P_2}{P_1} \right)^{\left(\frac{n-1}{n} \right)} - 1 \right]$$ (1.2)

where H_p = polytropic head, ft
         MW = molecular weight
         Z_{AVG} = average compressibility
         T_1 = suction temperature, °R
         n = compression coefficient
         P_1 = suction pressure, psia
         P_2 = discharge pressure, psia

One might increase the volume throughput by making the compressor larger. Or, one could speed up a compressor to increase its throughput volume. But there are interesting limits to these moves and Mach number is one of these. It is the ratio of two speeds and is thus a dimensionless number. Although not exactly as defined in physics, in a process gas compressor the *Mach number* is the speed of the gas moving through the machine at a particular location, divided by the speed of sound as it is in that process gas at its particular physical conditions, including those of temperature and pressure. Allowable Mach numbers differ with gas density and excessively high numbers adversely affect compressor performance.

Gas density is influenced by gas molecular weight and pressures. The maximum allowable velocity of a gas rushing through compressor impellers, stationary return bends, and nozzles is limited and is quite obviously a function of gas conditions. Moreover, elevating gas pressures tends to require thicker compressor walls than needed at low pressures. All of these constraints must be taken into account by the diligent designer.

All gas compressors, regardless of type, will deliver a compressed gas with an increased discharge temperature. Therefore, not all the energy put into the gas will be converted to increased pressure. The "practical" energy conversion is called *polytropic head*; it could be described as the intangible measurement of energy density imparted to a gas stream by the compressor. Adding this energy results in a pressure increase as the gas passes through the machine. Some head calculations assume (quite erroneously) that no heat is added to the gas. Calculations labeled "polytropic," meaning more versatile and affecting many kinds or types, are more realistic. In the polytropic method this energy density calculation takes into account heat being generated in the process. The polytropic head generated by a given impeller is a function of gas molecular weight, the thermal behavior—specific heat—of the gas, compressor efficiency, compressor inlet temperature, and compression ratio.[2]

In a centrifugal compressor the ratio of specific heats (c_v/c_p, the k-value) has marked influence on the design. Assume for the sake of illustration that a single impeller compresses alternately two gases of equal density, one having a k-value of 1.1, the other of 1.4. As compression takes place, temperature increases are different in gases with a 1.1 or 1.3 k-value versus a gas with a 1.4 k-value (Fig. 1.1). Consequently, the average density or specific gravity of the lower k-value gases will diminish less than that of the high k-value gas. The result is that the pressure generated by the single impeller under these identical conditions of peripheral speed and volume at inlet will be less for the high k-value gas than for the low k-value gas.[3]

In Eq. (1.2), the exponent n incorporates this factor k and we might just let it go with that. There is also a general rule of thumb that assumes that a single impeller will accommodate an unlimited mix of these parameters, up to 10,000 ft of head per impeller. If a calculation using Eq. (1.2) showed head between 10,000 and 20,000 ft one would require two impellers in series, or if it were between 20,000 and 30,000 ft, one would build a three-stage compressor. For somewhere between 80,000 and 90,000 ft, nine stages of compression would be needed, and so forth. Up to 11 stages of compression have

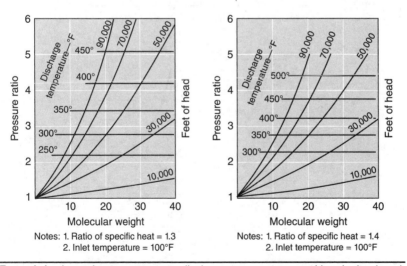

FIGURE 1.1 Approximate compressor discharge temperatures and heads developed at different pressure ratios and molecular weights. Note two different ratios of specific heat are considered: 1.3 (left illustration) and 1.4 (right illustration) (Ref. 1).

been accommodated in a single compressor casing. If more stages of compression are required, then another casing needs to be added. Some compressor trains have as many as five or even six bodies. A compressor train with an electric motor driver, a speed-up gear, two compressor bodies, another speed-up gear, and a third compressor casing would comprise six bodies.

Depending on thermal behavior or specific heat, and also depending on inlet temperature and compression ratio, the compressor discharge temperature may be too high for a particular metallurgy. In that instance, intercooling is often provided after typically three of four stages of compression. The compressed gas is then cooled and routed so as to reenter the compressor into the next section of upstream stages. (Note: For an excellent overview of the many available options consult Ref. 4.)

Positive Displacement versus Dynamic Compressors

Compressors are separated into positive displacement types and dynamic types. Reciprocating compressors are positive displacement machines whereas centrifugal and axial compressors are "dynamic" machines. All are involved in energy transfer and all require a measure of maintenance or similar upkeep. Their fields of application overlap and are shown in Fig. 1.2.

A few principal parts of reciprocating compressors are highlighted in Fig. 1.3. Because gas is compressed as the piston moves

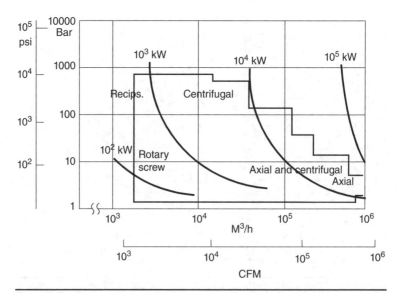

FIGURE 1.2 Compressor application ranges. Overlap of application ranges for different compressor types (Ref. 5).

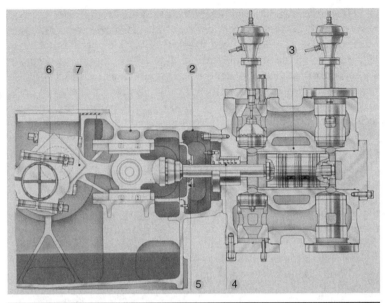

Figure 1.3 A double-acting reciprocating compressor schematic showing (1) the cooling water jacket in the general vicinity of the sliding "crosshead" mechanism; (2) a distance piece or similar compartment which allows access to the packing (4) and wiper ring (5) through which the piston rod travels; (3) a compressor cylinder with replaceable liner sleeve; (6) the bolting arrangement associated with one of the connecting rod bearings; (7) the rotating crankshaft.

back and forth in the cylinder, the machine is called double acting. Small air compressors are occasionally designed as single-acting machines, meaning that the gas (air, in this instance) is being compressed only as the piston advances. The cross-section of a single-acting compressor is simpler than the partial one shown in Fig. 1.3 for a double-acting process gas machine.

While there are many reciprocating or rotary positive displacement compressors used in the lower flow rate services, high flow rate process gas machines are typically centrifugal compressors. There are also axial compressors for the movement of large volumes of gases. All are a vital link in the conversion of raw materials into refined or finished products, and also for economically transferring energy from one form into another. Some compressors are used for the extraction of raw materials in mining operations, others are applied in conservation of energy by reinjecting natural gas, or in secondary recovery processes in crude oil-producing fields, or for oil recovery from shale and tar sand deposits. Compressors also deliver reaction air, oxygen, and other gases in virtually every chemical, gas processing and refining facility.

The economy and feasibility of the numerous compressor applications depend on high equipment reliability. A particular segment of our text will deal with the need to address reliable operation of auxiliary equipment, such as lubricating oil and seal support systems, or surge control devices. In general, selection takes into account the required gas throughput, desired efficiency, and the space taken up by a compressor—its "footprint." A good selection routine is heavily influenced by maintenance philosophies and the frequency of planned or budgeted shutdowns. In general, the less expensive machines require more frequent maintenance-repair-overhaul (MRO) shutdowns. One of the chapters on reciprocating compressors (see Chapters 12 through 15) will deal with the issue in greater detail.

The considerable overlap between machine types was noted in Fig. 1.2. There is overlap in throughput capacity or even in achievable pressures; one of the chapters on centrifugal compressors will explain that in somewhat greater detail.

Before contrasting the different types of compression machinery, we should define reciprocating compressors as machines where crankshaft rotation is transferred to one or more pistons that move back-and-forth in a cylinder or cylinders. Dynamic or turbocompressors are rotating machines with impellers (Figs. 1.4 through 1.6). They could also be rotating machines with all freestanding blades, or rotors with freestanding blades in the lower stages feeding gas to higher stage impellers in the same compressor casing (Fig. 1.7).

The casings of dynamic compressors are either horizontally split or vertically split. A six-stage horizontally split machine is depicted in Fig. 1.5. To gain access to the rotor, the entire upper half of the casing will have to be removed. As the designation implies, the split line between upper and lower casing halves is quite obviously horizontal. The nozzle orientation is downward and nozzles are part of the lower casing. If the nozzles had been provided on the upper casing, the connected piping would have to be dismantled before the upper casing could be lifted off for rotor maintenance. With nozzles on the lower casing, piping can remain in place while performing rotor maintenance; however, the entire turbocompressor will have to be mounted on an elevated platform ("mezzanine") to accommodate the downward nozzles and their associated pipe works.

Note the mirror-image orientation of the three lower stages on the right relative to the three higher stages on the left in this illustration. This is done for thrust balance.

An eight-stage vertically split turbocompressor is shown in Fig. 1.6. The casing is essentially a thick-walled pipe, and vertically split compressors are often called "barrel" machines. To gain access to the rotor, one or both of the massive end walls will have to be removed and the rotor and its surrounding diaphragms pulled out axially from the pipe-like casing. The designation "vertically split" is a bit misleading and is chosen to distinguish it from the horizontally

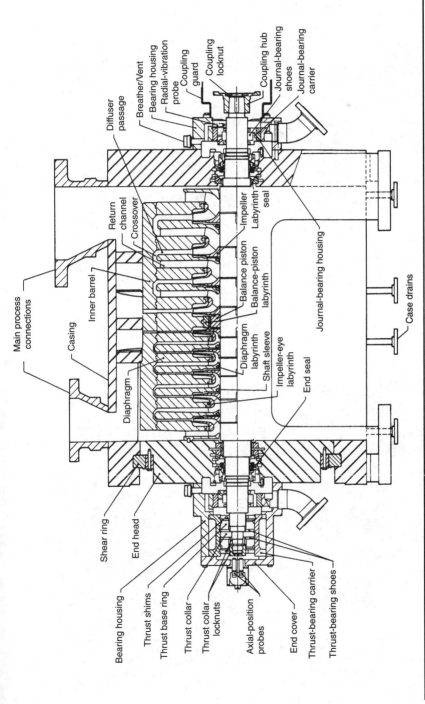

FIGURE 1.4 Centrifugal compressor nomenclature (per API-617, Ref. 6).

Main process connections

Casing

Inner barrel

Diaphragm

Shear ring

End head

Bearing housing

Thrust shims

Thrust base ring

Thrust collar

Thrust collar locknuts

Axial-position probes

End cover

Thrust-bearing carrier

Thrust-bearing shoes

Diffuser passage

Return channel

Crossover

Breather/Vent

Bearing housing

Radial-vibration probe

Coupling guard

Coupling locknut

Coupling hub

Journal-bearing shoes

Journal-bearing carrier

Impeller

Labyrinth seal

Balance piston

Balance-piston labyrinth

Journal-bearing housing

Diaphragm labyrinth

Shaft sleeve

Impeller-eye labyrinth

End seal

Case drains

9

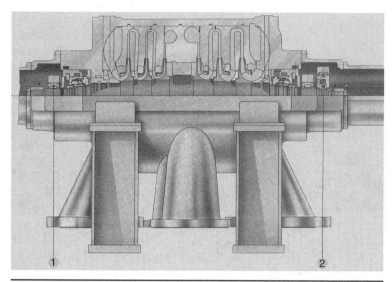

FIGURE 1.5   A six-stage horizontally split centrifugal compressor. The radial bearing (1) is shown on the left, the thrust bearing is located near the right in this particular illustration. (*Source:* Borsig GmbH, Berlin, Germany.)

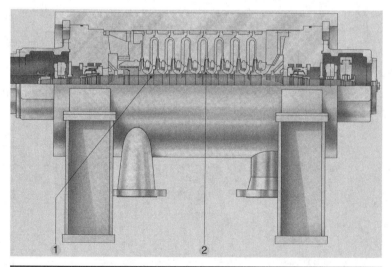

FIGURE 1.6   An eight-stage vertically-split "barrel-type" centrifugal compressor. Labyrinths (1) are located between the stages and all eight stages are oriented in the same direction in this particular illustration. An impeller or wheel (2) is often called a stage. (*Source:* Borsig GmbH, Berlin, Germany.)

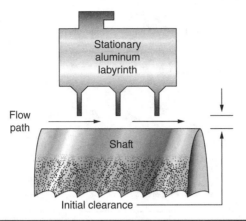

FIGURE 1.7   As-installed clearances allow relatively little parasitic flow (Ref. 7).

split counterpart machine. Internal labyrinth seals (1) made of a frangible material separate the individual stages in Fig. 1.6. [Note that internal labyrinths were also identified earlier (Fig. 1.4).]

Internal Labyrinth Seals

The internal labyrinths of modern centrifugal compressors are carefully designed and only material selection is more important than design.[7] The original or as-designed clearance in old-style aluminum labyrinths (Fig. 1.7) is lost after a rubbing event; even shaft damage is often noted (Fig. 1.8).

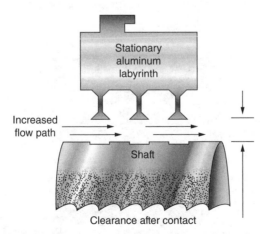

FIGURE 1.8   Flow path increases after rubbing contact and tends to damage both seal and shaft (Ref. 7).

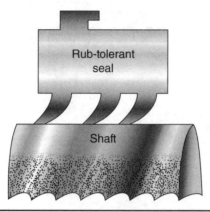

FIGURE 1.9 Rub-tolerant seals can deflect, but will generate heat upon prolonged contact. Although wear may be limited, these plastic parts are rarely suitable for long run-lengths expected of major process gas machines (Ref. 7).

Plastic sealing teeth configured to deflect and move out of the way are shown in Fig. 1.9. They are considered rub tolerant and can deflect. However, they do generate heat upon prolonged contact. Although wear may be limited, these plastic parts are rarely suitable for long run-lengths expected of major process gas machines.[7]

As a general rule, the seal combination of Fig. 1.10 deserves our attention. It incorporates best heat dissipation with labyrinth teeth integrated in the shaft and an abradable seal material located opposite it.[7] Finally, we show two examples of mounting an abradable material in a centrifugal compressor (Fig. 1.11).

All eight stages (labeled "2" in Fig. 1.6) are facing in the same direction. A balance piston near the entry to stage 4 (on the left) makes it possible to use a relatively small thrust bearing near the right side of this compressor.

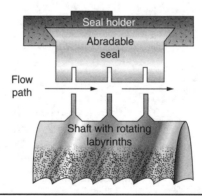

FIGURE 1.10 Best heat dissipation is achieved with labyrinth teeth on shaft and abradable seal material located opposite it (Ref. 7).

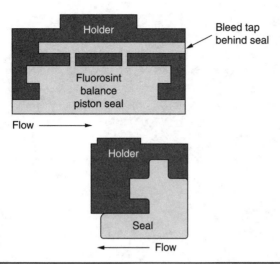

FIGURE 1.11 Acceptable mounting methods for abradable labyrinth materials (Ref. 7).

Horizontally split machines are generally used for lower discharge pressures; vertically split machines are needed at high pressure ratings. Nozzle orientation on vertically split "barrel" compressors is far less critical because maintenance rarely requires disturbing the piping. The nozzle locations may be varied to gain space needed for sidestream nozzles, or to accommodate nozzles needed for routing gas to and from external heat exchangers.

Vertically split compressors are sometimes placed on mezzanines to facilitate gravity drainage of lubricating oil and other fluids. This style of compressor casing is often placed at the end of a compressor train and free space is allocated to facilitate rotor removal.

Figure 1.12 shows a compressor rotor with seven axial (freestanding blade) and two centrifugal stages. Axial blading allows high volumes and moderate compression ratios; radial impellers can produce higher differential pressures. Combining the two approaches on a single rotor often represents an optimized design.

Turbocompressors handle any gas and almost any capacity. There are far more centrifugal compressors in service than axial compressors, although axial compressors are capable of the largest throughput volumes. Large process units, such as ammonia plants, ethylene plants, and base-load LNG plants became technically and economically feasible because of turbocompressor developments. In fact, an entire industry centers on the exploitation of hydrocarbons from oil and natural gas. By the mid to late 1900s this had resulted in hundreds of new processes that apply turbocompressors.

FIGURE 1.12   A rotor combining seven stages of axial blading with two radial compression stages. (*Source:* Demag, Duisburg, Germany.)

Preinvesting with Future Uprating in Mind

It is quite customary to purchase centrifugal compressors with future upgrading in mind. For increased throughput one might replace the very first impeller with a new, wider one. The present first impeller would be relocated to be in position 2 on the rotor, and so forth. The last impeller on the old (original) rotor would be discarded; in its place on the now new (or uprated) rotor would be the impeller which on the original rotor was next-to-last. However, so as not to exceed allowable Mach numbers in the compressor nozzles, the compressor casing would be purchased with future uprating in mind. The discharge nozzle, especially, might be specified to be somewhat larger than initially needed for the gas throughput originally specified. The incremental cost for preinvesting in larger nozzles versus specifying only the originally needed nozzles is generally quite low.

Reciprocating compressors are generally used for low flow volumes and/or relatively high discharge pressures.

What We Have Learned

The properties of the gases handled, the flow rates, the pressure differences to be overcome, and the operating temperatures can vary within very wide limits, depending on the type of service and production capacity.

Compressor selection takes all these into account but is not limited to the obvious. Among the lesser known factors are compressor locations and spare parts availability, climate factors that may affect cooling, availability of a trained workforce, toleration of shutdown duration and frequency, etc. These, together with failure analysis and troubleshooting, will be covered in some of the chapters in this text.

Internal labyrinths should be made of a frangible material. If rotor contact occurs, the material will fracture into relatively small pieces and pass through the machine. Metals used for labyrinth seals in older compressors often have a tendency to seize or gall.

Preinvesting in future uprate capability is easy to cost-justify.

References

1. Roots Compressors, Sales Literature, Connersville, IN, 1984.
2. Bloch, Heinz P., *Practical Guide to Compressor Technology*, John Wiley & Sons, Hoboken, NJ, 2006.
3. Alderson, William T., "Factors that influence selection of a compressor," *Chemical Engineering*, June 1956.
4. Bloch, Heinz P., and Arvind Godse, *Compressors and Modern Process Applications*, John Wiley & Sons, Hoboken, NJ, 2006.
5. Adapted from Sales Literature provided by Sulzer Turbo, Winterthur, Switzerland, 1981.
6. American Petroleum Institute, API Standard 617, Centrifugal Compressors, Alexandria, VA.
7. Quance, Steve, "Using plastic seals to improve compressor performance," *Turbomachinery International*, January/February 1997.

CHAPTER 2

Selection Factors for Process Compressors

E very style or type of compressor was developed to serve a particular purpose and the collective references have devoted hundreds of thousands of pages to the topic.[1,2] The overall performance trends of three prominent compressor styles are shown in Fig. 2.1. In each case, our interest is concentrated on a design point defined as the point where a desired 100 percent pressure performance meets 100 percent gas throughput performance. The illustration shows that as the pressure output of a positive displacement (primarily reciprocating and rotary screw style) compressor is increased, the volume being compressed decreases slightly. It can be reasoned that some of the pressurized gas slips back as "blow-by" from the discharge side to the suction side of a positive displacement machine.

Centrifugal and axial compressors are called "dynamic" compressors. They produce a pressure ratio, i.e., *head*—a pressure development capability as a function of speed and impeller diameter in centrifugal, or blade length in axial compressors. Again looking at the intersection of 100 percent pressure and 100 percent flow, any increase in pressure ratio or discharging the compressed gas into a region of higher pressure will follow the curve. For the *centrifugal* compressor with typical performance represented in Fig. 2.1 a maximum ratio of 115 percent cannot be exceeded; also, as pressures are increased from 100 to 115 percent, the flow-rate or compressor throughput will decline from previously 100 to perhaps 72 or 75 percent. An *axial* compressor will perform somewhere between the centrifugal and positive displacement machines; their generalized performance characteristics are plotted in Fig. 2.1. It can be said that these approximations are reasonably representative of what we find in process industry applications. However, the specifics can vary quite a bit.

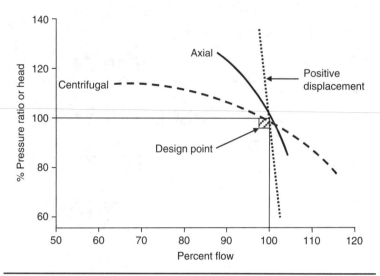

Figure 2.1   Selection factors: Compressor performance characteristics.

A relative operating flexibility overview is presented in Table 2.1 for the two dynamic compressor styles, centrifugal and axial, and the two positive displacement compressor styles, reciprocating and rotary. Five generally important attributes are listed and the letters E, G, F, and P inserted as the rankings excellent, good, fair, and poor.

Operating Flexibility	Axial	Cent	Recip	Rotary
1. Ability to accommodate multiple operating points at good efficiency	P	G	E	F
2. Ability to reduce volume flow at fixed speed*	F	F	E	G
3. Ability to reduce volume flow by speed control*	P	P	E	E
4. Ability to accommodate reduction in molecular weight*	F	F	E	G
5. Ability to accommodate increase in pressure ratio*	F	P	G	G

*With changes in speed or using control devices. All other operating parameters fixed.

Table 2.1   Selection Factors: Compressor Operating Flexibility Comparison

Minimizing Capital Costs	Axial	Cent	Recip	Rotary
1. Bare compressor, minimum controls	F	G	F	E
2. Controls for part load operation	F	F	G	G
3. Controls for parallel operation	F	P	G	G
4. Foundation and installation	E	G	P	E
5. Driver	G	G	P	E
6. Ability to accommodate side steam on one frame	P	E	E	P
7. Ability to accommodate different services on one frame	P	P	E	P

TABLE 2.2 Selection Factors: Compressor Capital Cost Overview

Whenever minimizing capital cost is of greatest importance, Table 2.2 attempts to give somewhat elementary guidance. It is fair to point out that cost-of-ownership (synonymous with life cycle cost) includes issues and considerations that go well beyond initial cost. In fact, initial cost can be less than 8 percent of overall cost. Operating and maintenance costs inevitably make up the bulk of total cost of ownership for process gas compressors. Reliability-focused users will not make compressor selection on initial cost considerations alone. They make sure that cost estimates reflect, and capital budgets relate to, only reliable equipment. It is a widely prevalent mistake to base one's project estimates on lowest initial cost of machines and then demand or expect high equipment reliability.

Highly reliable compressors will cost more than marginally reliable ones. Compressors with marginal reliability will either have to be shut down frequently for preventive maintenance or will fail catastrophically if needed maintenance work is deferred or omitted altogether. It may be possible to use condition monitoring instruments to obtain advance warning of component distress; however, it is doubtful if engaging in all manner of predictive maintenance (PdM) is more cost-effective than buying a more reliable compressor.

The operating environment of a compressor may affect the different compressors to a different degree. Likewise, it can be said that attributes and capabilities of certain compressors influence the environment which a particular compressor then creates.

Nine environmental considerations are tabulated in Table 2.3; note again that the relative categories or rankings E, G, F, and P are chosen from the authors' experience.

Environmental Considerations	Axial	Cent	Recip	Rotary Screw
1. Resistance to damage by entrained solids	P	F	G	E
2. Resistance to damage by entrained liquid droplets	P	F	G	E
3. Capability to deliver oil-free gas	E	E	E	E
4. Ability to limit discharge temperature by internal cooling	P	F	F	E
5. Freedom from vibration	E	E	P	G
6. Freedom from noise	F	F	G	P
7. Freedom from piping pulsations	E	G	P	F
8. Ability to compress dirty, wet gas	P	P	F	G
9. Gas tightness	G	E	G	G

TABLE 2.3   Selection Factors: Nine Environmental Considerations

Table 2.4 gives generalized maintenance cost rankings; three important listings make up the illustration and reciprocating process compressors are given a seemingly low ranking. However, and as compiled in Table 2.5, reciprocating compressors still have certain attributes that can place them at the top of competing machines.

One might increase the volume throughput by making the compressor larger. Or, one could speed up a compressor to increase its throughput volume. Elevating gas pressures tend to require thicker compressor walls than needed at low pressures.

Minimize Maintenance Cost	Axial	Cent	Recip	Rotary Screw
1. Number of items to be routinely inspected or replaced	G	E	P	E
2. Ability to operate reliably without outage	G	E	F	G
3. Length of time between scheduled overhauls	G	E	F	G

TABLE 2.4   Selection Factors: General Maintenance Cost Rankings

Efficiency, Power Required	Axial	Cent	Recip	Rotary Screw
1. Efficiency level at design point	E	G	G	F
2. Flatness of efficiency characteristic at constant speed, changing volume flow	P	G	E	F
3. Simplicity of indirect intercooling	P	E	E	P

TABLE 2.5   Selection Factors: Efficiency Favors Reciprocating Compressors

What We Have Learned

There are trade-offs among compressors and no one type or configuration fits all circumstances. The maintenance philosophies and uptime requirements of a particular project determine the selection, and many factors should be considered. All are based on experience.

Purchasing blindly on the basis of low bids and rapid delivery may compromise compressor reliability. All factors are important and must be given consideration before making procurement decisions.

References

1. Bloch, Heinz P., *A Practical Guide to Compressor Technology*, McGraw-Hill, New York, 1996; 2d ed., John Wiley & Sons, Hoboken, NJ, 2006.
2. Bloch, Heinz P., and Arvind Godse, *Compressors and Modern Process Applications*, John Wiley & Sons, Hoboken, NJ, 2006.

CHAPTER 3

Operating Characteristics of Turbocompressors

This text is not meant to duplicate the hundreds of other texts on compressor operation. Instead, its intended function is to synthesize and to address the needs of our target audience. As authors, we believe the operating characteristics of turbocompressors must be understood if safe and efficient operation is to be obtained.

The relationship between the inlet volume and the discharge pressure of a centrifugal compressor is called its characteristic curve; the terms characteristic curve and performance curve are used interchangeably for these curve-shaped representations (Fig. 3.1).

The discharge pressure developed by a centrifugal compressor is always equal to the total system resistance; it is this resistance (composed of valves and piping) that must be overcome. When the system resistance (or pressure to be overcome) changes, the rate of flow through the compressor will automatically adjust itself to equal the new resistance. This automatic process occurs over the entire operating range of the compressor, within its limitations.

There is a maximum volume that any compressor will deliver. As seen in Fig. 3.1, the characteristic curve tends to become vertical at maximum volume. At this point, called "stonewall," flow cannot increase despite a great reduction in system resistance.

With decreasing flow and toward the left side of the performance curves, there is a minimum point called "surge" (or surge limit) below which unstable operation will result.

Surge

Surge represents an unstable region of flow (Fig. 3.1). All centrifugal compressors have a flow at which maximum pressure generation is possible. Further reduction in flow beyond this point results in a

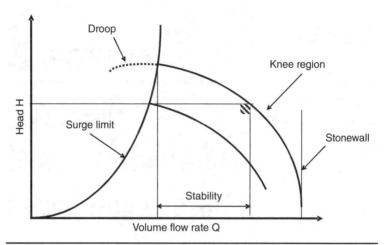

Figure 3.1 Operation: Compressor characteristic curve and typical curve terminology (Ref. 1).

decreasing discharge pressure. When this occurs, the pressure in the system exceeds that produced by the compressor and momentary reversal of flow occurs. As this momentary flow reversal reduces the system pressure, the compressor will rapidly build up pressure. With pressure regained, flow will again proceed in the normal direction. Oscillation back and forth (surge) will continue until the system resistance, or discharge pressure, is reduced. Surge cycles can have durations from 0.05 seconds to 5 or more seconds. All kinds of variables are at work and, while some machines survive days of surge, some will fail after less than 60 seconds of severe surge.

If process flow is less than the compressor surge, the compressor should be operated in a stable region and excess flow blown off or bypassed (recycled) through a fast-acting surge control valve. For air applications, this excess can be blown off to atmosphere. For process gas applications where the gas cannot be lost, the excess can be recycled or bypassed through a cooler (also called a heat exchanger) back to the compressor inlet. The surge control valve can be operated either manually or with automatic controls. Quite obviously, fast-acting automatic controls are much preferred on valuable process gas compressors.

Compressor Speed

The characteristic curve is fixed for a constant speed. For a variable speed driver, such as a turbine, the curve location will vary with a change in speed—see lower curve in Fig. 3.2. However, the "new" curve is essentially parallel to the upper curves. Speed variation is an efficient means of compressor control to meet several operating flows and pressures. With a fixed system resistance, flow will vary directly

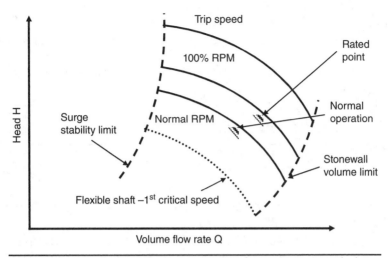

FIGURE 3.2 Operation: Centrifugal compressor performance curve terminology (Ref. 1).

with speed, while head varies as the square of the speed, and horse-power as the cube of the speed (Table 3.1).

Centrifugal and axial compressor performance changes can be estimated from the fan laws, also called affinity laws (Table 3.1). A numerical example is used in Table 3.2.

Changes in RPM (revolutions per minute) are most often used to adjust compressor performance, obtain pressure rise, modify through-put, etc. We can observe how speed changes affect other parameters in Fig. 3.3. If, as an example, the speed (RPM) is increased from previously 100 percent to now 110 percent of design, the discharge temperature (T_d), head developed (H), power demand (HP), weight flow, volumetric throughput, and the input power demand will all increase in different proportions.

- Centrifugal and axial compressor performance can be estimated for points other than rated flow and speed.
- Use between 85% and 100% of rated rotational speed.
- Since changing impeller blade outer diameter also changes the peripheral speed, the diameter can be substituted for RPM.
- $Q_2/Q_1 = \text{RPM}_2/\text{RPM}_1$.
- $H_2/H_1 = (\text{RPM}_2/\text{RPM}_1)^2$ (for constant Q).
- $\text{HP}_{2\,gas}/\text{HP}_{1\,gas} = (\text{RPM}_2/\text{RPM}_1)^3$ (assumes head and flow allowed to rise).

TABLE 3.1 Operation: Fan laws (Affinity Laws) Pertaining to Centrifugal and Axial Compressor Performance Curve Changes (Ref. 1)

Example of its use:

Let's say the RPM was increased by 10% and we wanted to know what the horsepower increase would be:

- $HP_{2\,gas}/HP_{1\,gas} = (RPM_2/RPM_1)^3$ (assumes head and flow allowed to rise).
- $HP_{2\,gas} = HP_{1\,gas} \times (RPM_2/RPM_1)^3$.
- $HP_{2\,gas} = HP_{1\,gas} \times (1.10)^3 = 1.33 \times HP_{1\,gas}$.
- The gas horsepower would thus increase by about 33%.

TABLE **3.2**   Operation: Numerical Example of Fan Laws (Affinity Laws) Pertaining to Centrifugal and Axial Compressor Performance Curve Changes (Ref. 1)

Inlet Pressure and Temperature

The characteristic curve (Fig. 3.3) of a given compressor has a fixed pressure ratio (absolute discharge pressure divided by absolute inlet pressure) at any given flow. The back pressure in this illustration is uncontrolled. If inlet pressure is reduced, the discharge pressure will be reduced equal to the new inlet pressure times the pressure ratio. In addition, the horsepower will vary directly with inlet pressure, and will decrease proportionally. With an increased inlet pressure, both discharge pressure and horsepower will increase.

An increase in inlet temperature will result in an opposite reaction to the discharge pressure and horsepower; it will cause a lowering of these two parameters. Conversely, decreased temperature will

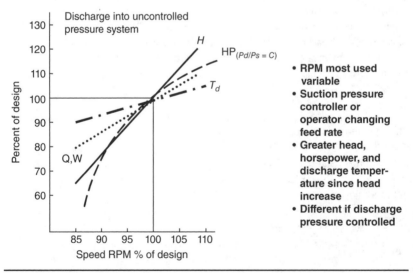

FIGURE **3.3**   Operation: The effect of speed changes on other compressor performance parameters (Ref. 1).

raise pressure and horsepower, because mass flow is increased. Whenever an unduly low temperature is reached, motor overload could occur.

The compressor performance curves are developed by the manufacturer and follow Eq. 1.2. Needed are:

P_1 = intake pressure, psia

T_1 = intake temperature degrees F absolute (= deg. F + 460) or Rankine temperature

V_1 = intake volume, icfm, measured at intake temperature and pressure

W = weight flow, expressed as lb/min.

P_2 = discharge pressure, psia; $k = C_p/C_v$ of gas

1545 = universal gas constant (molar system); MW = molecular weight of gas

Z_1 = compressibility factor at intake pressure and temperature

Z_2 = compressibility factor at discharge pressure and temperature

$(Z_1 + Z_2)/2$ = average change in compressibility during compression from P_1 to P_2

R = ratio of compression, P_2/P_1

The weight flow can be converted to volume flow rate (Q) by using the formula:

$Q = (W) \times (10.729) \times (T_1) \times (Z_1)/(P_1) \times (MW)$—actual cubic feet per minute, ACFM

Effect of Gas Properties

Changes in gas mixture properties will affect the compressor characteristic curve. For a given inlet flow, a decrease in the molecular weight will reduce both the discharge pressure and horsepower. An increase in the molecular weight will result in an increase in both the pressure and horsepower.

A decrease in the ratio of specific heat will increase both discharge pressure and horsepower. Conversely, an increase in k-value produces a decrease in pressure and horsepower. This is summarized in Table 3.3.

Effect of Speed Changes

Recall that the back pressure in Fig. 3.3 was left uncontrolled. If the discharge pressure is controlled by a back pressure valve, the head will remain constant because suction and discharge pressures remain constant.

Increase	Results
Suction temperature	Decreased discharge pressure Decreased mass flow capability
Suction pressure	Increased discharge pressure Increased mass flow capability
Molecular weight	Increased discharge pressure Increased mass flow capability
Specific heat ratio	Decreased discharge pressure Approximately same mass flow

TABLE **3.3**   Effect of Changing a Single Gas Condition on Other Parameters

As speed is increased, the power demand will increase as the cube of speed changes; weight and volume flow will increase in the same proportions as the power input.

Changing Suction Temperature

The effects of changing suction temperature were given in Table 3.3 and are restated in Fig. 3.4. An increase in suction temperature increases the head required to generate pressure. Compressor speed to supply head increases and more power is needed. Suction flow increases, but because of lower density the resulting weight flow decreases. The discharge temperature also increases. In essence:

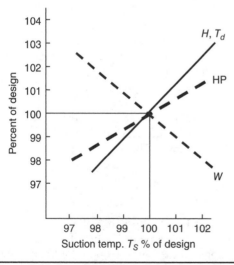

- Increase in suction temperature results in increase in head required.
- Compressor speed increases to supply head, which requires more horsepower.
- Suction flow increases but because of density effect weight flow decreases and discharge temperature increases too.

FIGURE **3.4**   Operation: Effect of changing suction temperature (Ref. 1).

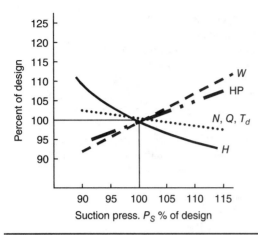

• **Suction pressure is the most important process variable.**
• Since pressure is energy, the required head *H* is reduced as is the speed and volume flow.
• Increasing pressure increases the gas density and thus the weight flow *W* increases the throughput, but requires more horsepower.

Figure 3.5 Operation: How changes in suction pressure affect other parameters (Ref. 1).

• **Droop** – The drop in pressure at the outlet of a pressure regulator, when a demand for gas occurs.
• **Knee** – Slope change rapidly before "stonewall." Operation inefficient, erratic, seldom useful.
• **Head, polytropic** – The energy, in foot pounds, required to compress polytropically to deliver one pound of a given gas from one pressure level to another.
• **Performance curve** – Usually a plot of discharge pressure versus inlet capacity and shaft horsepower versus inlet capacity.
• **Rise** – Must rise slightly so compressor will operate stably.
• **Stability** – Means stabile flow range between surge and normal flow.
• **Stonewall** – Vertical portion curve, defines maximum flow rate compressor canpass. As approach sonic speed turbulence, shockwaves, flow separation occur.
• **Surge** – A phenomenon in centrifugal compressors where a reduced flow rate results in a flow reversal and unstable operation.
• **Surge limit** – The capacity in a dynamic compressor below which operation becomes unstable.
• **Temperature, absolute** – The temperature of air or gas measured from absolute zero. It is the Fahrenheit temperature plus 459.6 and is known as the Rankine temperature. In the metric system, the absolute temperature is the Centigrade temperature plus 273 and is known as the Kelvin temperature.
• **Turndown** – Mass flow reduction possible before encountering surge.

Table 3.4 Operation: Some Terms Often Found on Compressor Performance Plots (Ref. 1)

Keeping suction temperatures as low as possible is important if compressor throughput (weight flow) is to be maximized.

Finally, the effects of changes in suction pressure can be visualized from Fig. 3.5 and frequently used terms summarized in Fig. 3.6. The x-axis in Fig. 3.5 displays suction pressure percentage change relative to the original design. Assuming we wanted to maintain a certain discharge pressure, increasing the suction pressure causes the gas density and throughput mass (weight flow) to increase. The required head H, speed N, volume flow Q, and discharge temperature T_d will decrease. On the downside, energy demand will increase as the weight flow increases.

What We Have Learned

Experience tells us that surge must be avoided; to avoid surge one installs surge abatement measures. Surge abatement is typically achieved by rapidly opening a large bypass valve in a pipe loop from compressor discharge back to suction. This ensures that gas flow through the compressor stays to the right of the surge limit in Fig. 3.1.

Operator training must cover compressor performance and how changes in one parameter affect other parameters of interest. In the age of computerization and easy simulation, the various parameters deserve to be displayed in the operator's control room; showing them in dynamic, animated form is best.

Reference

1. Sofronas, Anthony, "Introduction to centrifugal compressor systems, centrifugal compressor operations for 21st century users," Continuing Education Short Course, The Turbomachinery Laboratory, April 25–29, 2005.

CHAPTER 4

Wet and Dry Gas Seals for Centrifugal Compressors

S haft seals keep the compressed gas from leaking out and the need for leakage prevention may include process economy, safety, and environmental protection. Moreover, the process gas will usually have to be kept away from the compressor bearings. In essence, seals are located between the compressed gas and the bearings; a number of different sealing arrangements and configurations are found in industry. An overview of four principal seal types is provided in Fig. 4.1. Within each of the four principal types one finds numerous variants.

Some wet face seals incorporate a stationary, a rotating, and a floating ring; others incorporate only a rotating ring clamped to the shaft and a spring-loaded nonrotating face with the springs anchored in a housing assembly.

Dry gas seals are mechanical face-type seals available as single, double, and tandem models. However, dry gas seals should not be confused with the internal labyrinth seals that separate individual stages inside a compressor casing (see also Chapter 1, Figs. 1.4 and 1.6).

For over five decades, centrifugal compressors have benefited from face and bushing-type mechanical seal technology. The early compressors of the 1950s often struggled with cumbersome labyrinth sealing and gas eductor arrangements. But, in the 1960s, many labyrinth configurations were rapidly replaced by a variety of liquid-lubricated seals that introduced small amounts of seal oil between either the sealing faces or the small gaps between stationary and floating sealing bushings. American Petroleum Industry Standard 617 has for many years outlined different sealing arrangements and "generic" dry gas seals are

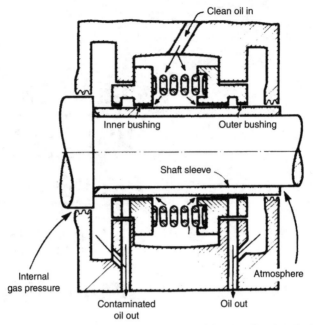

Clean oil in

Inner bushing · Outer bushing

Shaft sleeve

Internal gas pressure

Atmosphere

Contaminated oil out

Oil out

Courtesy of American Petroleum Institute, Standard API 617

Floating ring or bushing seal (1)

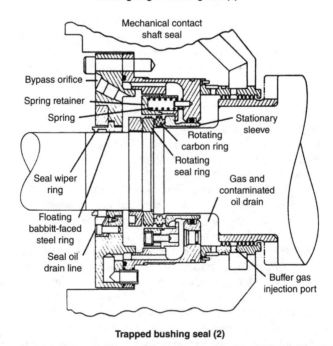

Mechanical contact shaft seal

Bypass orifice

Spring retainer

Spring

Stationary sleeve

Rotating carbon ring

Rotating seal ring

Seal wiper ring

Gas and contaminated oil drain

Floating babbitt-faced steel ring

Seal oil drain line

Buffer gas injection port

Trapped bushing seal (2)

FIGURE 4.1 Compressor seal overview depicting a floating ring or bushing seal (1), trapped bushing seal (2), liquid-lubricated mechanical contact seal (3), and dry gas face seal (4) (Ref. 1).

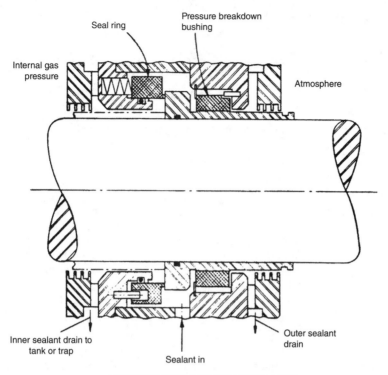

Liquid-lubricated mechanical contact seal (3)

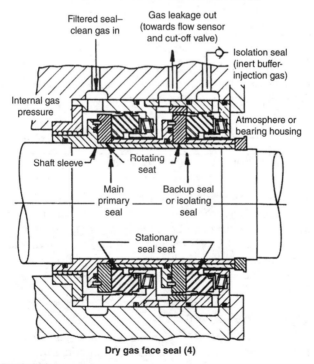

Dry gas face seal (4)

FIGURE 4.1   (Continued)

33

Experience-Based Recommendations for Compressor Seal Selection			
Application	**Service**	**Inlet Pressure[a] kPa (psia)**	**Seal Type**
Air compressor	Atmospheric air	Any	Labyrinth
Gas compressor	Noncorrosive Nonhazardous Nonfouling Low value	Any	Labyrinth
Gas compressor[b]	Noncorrosive or corrosive Nonhazardous or hazardous Nonfouling or fouling	69–172[c] (10–20)	Labyrinth with injection and/or ejection using gas being compressed as motive gas
Gas compressor	Noncorrosive Nonhazardous or hazardous Nonfouling	≥25,000 ≥3,600	Gas Seal[d] Tandem preferred
Gas compressor	Corrosive[e] Nonhazardous or hazardous Fouling	Any[f]	Oil seal, double gas[f]

[a]Operating seal pressure range.
[b]Where some gas loss or air induction is tolerable.
[c]Pressure ranges shown for labyrinth seals are conservative. Manufacturers extend this range upward, resulting in a debit due to power losses.
[d]Within state of the art.
[e]H$_2$S is the most common corrosive.
[f]Dry running gas seals often have pressure limitations below those of oil seals.

TABLE 4.1   Turbocompressor Seal Selection Guidelines

included in this important and well-established API-617 compressor standard. User preferences differ, but each type of seal has its application range, as shown in Table 4.1.[1] In general, dry gas seals are favored in new installations. On existing installations, wet (liquid) seals merit replacing only if they are troublesome and if good experience references are available for dry gas seals in the same service.

The Case against Wet Seals

In 2010, a consulting engineer summarized his experience; it apparently favored dry gas seals in a wide range of applications. For wet seals he reported the following:[2]

1. High seal oil consumption, even when new; typically about 5 gpd (gallons per day) per seal, for a total of 10 gpd per typical compressor. After 1 year in service, this will average about 15 to 20 gpd. For combined lube-seal systems (not always possible), the leakage losses are reduced, but a total of at least 2 gpd are lost into the compressor itself on an average-size machine. (Contaminated lube oil may damage compressor bearings, leading to vibration and downtime.)

2. Wet seal systems require more operator attention (man-hours of labor) because the system's sour seal oil drainage must be physically inspected and measured on combined systems and there are many active components in the seal oil system.

3. Incremental energy requirements due to parasitic losses on wet seals are typically ~30 HP (total) on average-size 10,000 hp compressors. Adding seal oil pump energy (~35 HP) will increase the total energy increment to 65 HP.

4. Hydraulic surges are a potential hazard on wet seal systems, some more so than others. This can lead to seal face damage and reduced component life. (Manufacturers rarely mention this risk and seldom size the oil accumulators properly. Even then, bladder-type accumulators tend to become frequent maintenance items—partly due to internal gas leakage and elastomer failures.)

5. Life of wet seal bushing is 2 to 4 years on average; however, some wet bushing seals reach 6 years with well-designed late-generation components.

6. Because seal bushings do wear (microscopic abrasive particles can accumulate at the bushing seal edge and grind the shaft mechanically), experience points to the advisability of precautionary repairs on every second seal intervention. At that time the complete rotor must be removed to repair damaged shaft regions, even though the shaft is hardened. (It should be noted that late-generation Isocarbon® liquid film face seals do not have this problem.)

7. With some wet seal systems there are four pumps to inspect and maintain on each compressor lube/seal oil skid; two are seal oil and two are lube oil pumps.

8. Significant gas leakage losses to flare from the sour seal oil traps may have to be considered. Unless a flare gas recovery unit is available or the gas is reintroduced at the compressor suction, gas losses can be as high as 30 times the equivalent dry gas seal loss. Up to 75 scfm per seal have been reported for a 10,000 hp compressor in average condition.

9. With wet seals considerable instrumentation complications exist because many control devices are needed. The cost of instrument maintenance can be high.

10. Troubleshooting oil seals requires a highly qualified engineering or senior technical staff. Experience shows that seal oil system design/installation/maintenance problems are not always resolved and constant vigilance is required. (Reference 2 concludes that gas seals require lesser skills.)

11. Contamination of the process gas path, such as heat exchangers and catalyst beds downstream of the compressor is often experienced. This can become a burdensome cost to the plant. On a per-compressor basis the yearly reduction in efficiency of heat exchangers on a propane system can easily incur $50,000 to $100,000 (year 2010 valuation) due to the resulting cooling losses.

12. Float traps on seal oil systems can become a maintenance issue.

13. Faulty level controllers can risk worrisome gas leakage to atmosphere.

14. For compressors equipped with bushing-type seals, no backup seal is available in case of power outage (loss of seal oil). As a consequence, emergency power is required for safe operation or an immediate shutdown is required. It should be noted that Ref. 3 states that dry seals have a safe record in this regard due to the availability of secondary sealing.

15. The plot plan ("footprint") needed by wet seal auxiliary systems is usually relatively large. The compressor skid itself is physically smaller and less expensive for machines equipped with gas seal systems.

How Gas Seals Function

There are many functional similarities between gas seals and their predecessors. These predecessors include many variants of face, bushing, and floating ring seals. Yet, there are also features that differ. For instance, the seal face of the rotating mating ring can be divided into a grooved area at the high-pressure side and a dam area at the low-pressure side (Fig. 4.2). The shallow grooves are often laser etched, spark eroded, or chemically milled. A typical depth is ~0.0003 in = 0.008 μ, quite obviously achieved through highly precise machining operations. T-shaped, V-shaped (bi-directional), and L-shaped (unidirectional) grooves have been produced; each configuration has its advantages and disadvantages. A stationary sliding ring is pressed axially against the mating ring by both spring forces and sealing pressure.

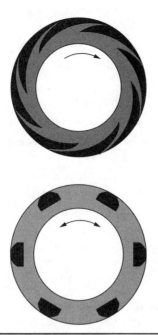

FIGURE 4.2   Mating ring vane-like grooves (top), U-grooves (bottom). Arrows indicate sense of rotation (Ref. 4).

The sealing gap is located between the mating ring and the sliding ring. For proper noncontacting operation, these two rings have to be separated by a gas film acting against the closing forces in the sealing gap. The gas film is achieved by the pumping action of the grooves and the throttling effect of the sealing dam. Groove geometry is critical for trouble-free operation of the seal.

Minimizing the Risk of Sealing Problems

Dry gas sealing is obviously coming into prominence and deserves consideration. However, specification, review, purchasing, and installation of a dry seal support system cannot be left to chance. A thorough review of the owner's facility and the particular process unit in which the compressor will be installed may well be the key to minimizing gas seal failure events. In order of importance, the following factors should be considered in examining dry seal support systems for centrifugal compressors:

Gas Composition: Understanding the actual gas composition and true operating condition is essential, yet often overlooked. For example, it is necessary to understand when and where phase changes start and that condensed liquids must not be allowed in the sealing gas.

Commissioning procedures and control system design must be thoroughly understood:

1. Is clean and dry buffer gas available at all anticipated compressor speeds?
2. Is the seal protected from bearing oil?
3. How is the compressor pressurized or depressurized?
4. How is the machine brought up to operating speed and how will the seal react?
5. Are all operating and maintenance personnel fully familiar with the compressor maintenance and operating manual?
6. Is the full control system included and adequately described in these write-ups?
7. Are the key elements of the system design understood and do they include buffer gas conditioning, heating, filtration, regulation (flow vs. pressure), and monitoring?

Before opting for dry gas seals in retrofit situations, ask if the apparently flawed liquid seals really represent the best the vendor was able to offer. The claims of competing vendors and claims made for different styles of seals must be checked against actual experience. In some cases these checks lead to the purchase of late-generation liquid film face seals instead of gas seals. The areas of safety and reliability must always be given special consideration.

Seal Safety and Reliability

Positive sealing of the compressor during emergencies must always be assessed. In the event that the gas pressure in the compressor casing exceeds that of the seal oil pressure, shutdown pistons in advanced face seal assemblies exert a force proportional to the gas pressure to keep the seal faces closed and prevent gas release to the environment.

Dry gas seals may or may not provide positive gas sealing if the seal faces are damaged or distorted. The backup seal may show reasonable performance under low pressure but may fail to perform under higher pressure if the primary seal fails.

The alarm and shutdown devices of the seal oil system must be of reasonable range and sensitivity and should provide reliable alarm and shutdown characteristics. Dry gas seal systems often rely on pressure switches of very low range and high sensitivity. They might have the tendency to malfunction or give a false sense of security. Make sure the system will provide a sufficiently high degree of alarm and shutdown performance.

Some seal configurations excel at online monitoring. The favored seal must allow easy verification of sound working condition. Beware of seal systems that are hampered by small diameter orifices that tend to plug; avoid sensitive pressure switches that often become inoperable. Dry gas seal failures are not as easy to detect as wet seal failures.

As a final comment, the user must go through a rigorous cost justification analysis. In many cases, gas seals are a good choice only if you do not have to purchase the oil seal console. If this console already exists, it's often difficult to justify gas seals.

With hydrocarbon gas prices escalating, it might be prudent to consider gas leakage rates in order to justify a conversion to gas-lubricated seals. Also not to be overlooked are advancements in traditional oil sealing technology for centrifugal compressors. Many world-scale manufacturers are now marketing improved versions of the oil seals that were originally furnished with their compressors.

Along these lines, one researcher suggests to study the failure statistics of your oil type seals and compare them with this: Current estimates of failure rates of gas seals are in the neighborhood of 0.175 failures/year, meaning that we could expect a problem every 6 years or so. At least one dry gas seal manufacturer bases recommended maintenance intervals around gas seals on limits set by the elastomer aging process. This manufacturer suggests the following maintenance routine after 60 months of operation:

- Replace all elastomers
- Replace the springs
- Replace all seal faces and seats
- Carry out a static and dynamic test run on a test rig

Making Good Choices

Thus, our final advice to the reader is to make informed choices. Consider gas seals only in conjunction with a clean gas supply. If your process gas causes the fouling deposits as shown in a later chapter (Figs. 12.2 and 12.4), ask critical questions of anyone offering seals for use with that kind of gas. If extensive microfiltration is needed, factor-in the cost of maintaining a dry gas seal support system. Look for seals that will survive a reasonable amount of compressor surging. Consider dry gas seals that incorporate features ensuring start-up and acceleration to operating speed without allowing the two faces to make contact. If these seals are not available from your supplier, look beyond the usual sources.

Some innovative manufacturers are offering modern dry gas seals for OEM as well as aftermarket applications. At least one manufacturer of advanced sealing devices has the capability to repair and test dry gas seals made by others.[5] Investigate the extent to which they can meet all of your reliability requirements; then consider using their dry gas seals.

What We Have Learned

Sealing technology merits a close review. Insist on references and check with other users on their experiences. Let all comparisons be fair and unbiased. Understand how seals function and accept only established facts. Discount anecdotal references and mere sales talk.

When using oil seals, ask where the excess oil will end up. Understand oil consumption requirements. Understand cost. Understand if startup conditions might allow entrained vapors to condense and ruin the seal.

When using dry gas seals, thoroughly understand gas purity requirements. Only pure gas can be used. Factor-in the cost of servicing auxiliary support equipment. Understand where the motive gas will go and what its operating costs are.

Regardless of seal type and style selected, the design of seal support systems should be under the compressor vendor's jurisdiction and the compressor manufacturer's warranty coverage. Resist the temptation to involve and deal separately with different suppliers.

However, seals and seal systems must be serviceable by your own workforces. It would not be prudent to depend entirely on the support of manufacturers and outside vendors.

References

1. Borsig GmbH, Sales Literature, Berlin, Germany, 1987.
2. Mid-East Turbomachinery Consulting Ltd., Personal reviews and professional communications with the co-authors. Dahran, KSA.
3. Bloch, Heinz P., and F.K. Geitner, *An Introduction to Machinery Reliability Assessment*, 2d ed., Gulf Publishing Company, Houston, TX, 1994, pp. 242–247.
4. Bloch, Heinz P., "Consider dry gas seals for centrifugal compressors," *Hydrocarbon Processing*, January 2005.
5. Carmody, Chris, "Dry gas seal repair and testing," *CompressorTech2*, December 2009.

CHAPTER 5

Bearings, Stability, and Vibration Guidance

O ver the years, compressor manufacturers have furnished different types of hydrodynamic bearings, ranging from plain sleeve bearings and pressure dam sleeve bearings to very precisely machined tilting shoe (tilting pad) bearings. Compressor shaft systems are supported by these bearings and some of these bearing styles incorporate fixed geometries (Fig. 5.1). Many incorporate grooves or dams so as to achieve the desired support stiffness or damping characteristics. An analogy comes to mind with tuned shock absorbers in automobiles. These provide damping to vehicle motions that cannot be suppressed by spring supports alone. In compressors, one wants successive vibratory excursions to decline rather than increase and amplify. In stable systems there will be a logarithmic decrement, whereas a vibration increment (an amplification) would be noted in unstable systems.

Mechanical Design

Although plain sleeve bearings are often quite satisfactory, they have, collectively, been somewhat susceptible to subsynchronous oil whirl and also high-frequency oil whip—all problems creating rotor instability. Whirling is orbiting of the rotor center in the bearing. The orbiting frequency is generally 0.42 to 0.47 times the running speed. Axially grooved bearings reduce instability risk, but require more oil flow to feed all grooves. They also have slightly reduced load-carrying capacity compared to full sleeve bearings. However, axially grooved bearings normally run cooler than plain bearings.

Lobe bearings with symmetrical and asymmetrical bore patterns represent an improvement over axially grooved cylindrical bore bearings. Their noncircular bores allow the oil film to form a wedge.

41

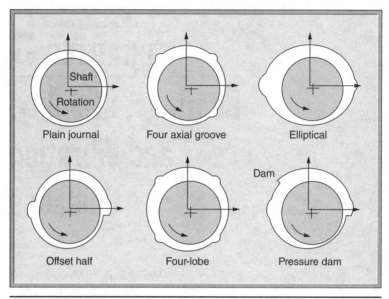

Figure 5.1 Some fixed geometry radial bearing designs (Ref. 1).

They were precursors to the present-generation tilting pad bearings. Tilting pads, also called tilting shoe bearings, have largely overcome the whirl and whip problems and are favored due to their ability to not only support, but also overcome small amounts of line-bore inaccuracies at the bearing housing locations of compressor casings. Tilting pad bearings are available in different geometries, with different numbers of pads, pad pivot locations, pad curvature, and different axial-width-to-bore-diameter ratios. Which one to use can only be determined by a rotor dynamics study.

While it would be technically desirable to have the pivot point of the bearing pads off-center and with the leading edge portion somewhat larger than the trailing edge portion to enhance stability, the resulting bearing would be unidirectional. However, centered pivots are found to predominate in tilting pad bearings since the bearing is now suitable for shaft rotation in either direction. Regardless of overall geometry, most hydrodynamic bearings with steel backing have babbitt lining of about 0.8 mm (0.032 in) thickness, and clearances of 1.5 to 2.0×10^{-3} in/in (mm/mm) of journal diameter. They utilize oil supply pressures in the relatively low range of 25 to 35 psi. Shaft rotation and bearing geometry cause higher pressures to be self-generated within the bearing. An oil film varying in thickness from 0.0001 to 0.001 in prevents metal-to-metal contact. Since shear action on the oil produces heat, the lubricant must be cooled. Circulating oil systems are best suited to accomplish the necessary cooling and filtering.

Destabilizing Forces and Aerodynamic Cross Coupling

Centrifugal compressors are custom-designed for particular gas properties and conditions. They are then shop-tested to verify proper mechanical operation. If, as an example, the gas molecular weight later changes at the process plant, there may occur performance changes in the compressor and the machine may exhibit fractional frequency vibrations of unacceptable magnitude.

Compressor performance is very much influenced by bearing design and gas loads acting on impellers. The bearing span (length of rotor) and shaft diameters may have been originally bought for operating conditions (MW, RPM) which were only slightly different from the conditions for which similar machines had been built before. Also, the original rotor was proof-tested in the machine before it was released by the equipment manufacturer.

Suppose a new compressor rotor was installed during an inspection and repair downtime (IRD) and would now operate under considerably different load and speed conditions. Assume also that the rotor could not be proof-tested in its casing because running on air would have resulted in excessive discharge temperatures and running on the process gas would be costly and time consuming.

Basically, then, the user would deal with the first compressor combining this particular bearing span, shaft diameter, gas load, and RPM and, running into a problem is always a possibility when you break new ground. So, when a problem first surfaces, reliability-focused users lose no time in identifying the root cause. The user can establish if a rotor vibrates at a frequency equal to its first natural frequency (critical speed) whenever a certain combination of speed, MW, and gas loading causes destabilizing forces in excess of the damping which normally inhibits vibrations in rotating machinery.

In many machines, these destabilizing forces are called aerodynamic cross-coupling.

These are simply the gas impulses which leave at the impeller vane exit edge. A simplified formula (Eq. 5.1) describes these forces as:

$$\text{Aero C-C} = [(K)(HP)(MW)(d_{exit})]/[(RPM)(d_{inlet})(D)(W)] \quad (5.1)$$

where
K = constant
D = shaft diameter
W = impeller exit width
HP = gas horsepower

d_{exit} = gas density, leaving
d_{inlet} = gas density, entering
RPM = speed of rotor
HW = molecular weight

Looking at this expression, we can see how there are many variables that tend to influence the machine's behavior. If the impeller exit width were to increase because operations have, say, flushed-off some polymer deposits, the factor Aero C-C is reduced and the machine vibration decreases. Next time when flushing is attempted, operations will be less successful because d_{exit} may have increased for some reason.

Fractional Frequency Activity in Bearings

Unlike once-per-revolution or twice-per-revolution frequency vibrations, fractional frequency vibrations are highly unpredictable and can get out of control very rapidly. That's why it is prudent to put certain machines on automatic trip (shutdown) whenever a predetermined value is reached. Perhaps this predetermined value should equal 50 percent of the bearing clearance, or should be governed by industry experience. One rule of thumb allows the vibration amplitude of the largest fractional frequency (or "subharmonic") to reach 15 percent of the amplitude at RPM frequency. If the amplitude climbs above 15 percent, the user's reliability professionals would recommend shutdown and repair.

Fractional frequencies and mysterious high frequencies are occasionally found in high-pressure compressors. Complex aerodynamics are often involved and anti-swirl vanes or destabilizing cascades (labyrinths) must be installed (Fig. 5.2). The intent is to dampen (meaning to obtain an amplitude decrement), and not to amplify (or increase) the vibration excursions. This is indicated in Fig. 5.3.

A top industry consultant once estimated that centrifugal compressors must be shut down in 3 to 10 seconds when the whirl orbit reaches 30 percent of the bearing clearance. Note that bearing clearance calculations are given under the head "Bearing Clearances." Suppose the bearing clearance is 0.008 in. Assume, for the sake of illustration, that the vibration amplitude reaches 3.2 mil (0.0032 in).

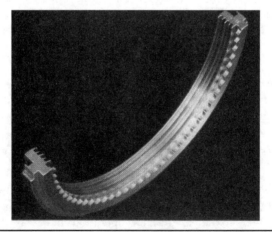

FIGURE 5.2   Anti-swirl cascade labyrinth (Ref. 2).

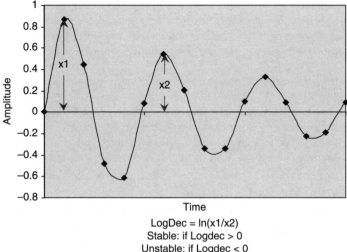

$$\text{LogDec} = \ln(x1/x2)$$
Stable: if Logdec > 0
Unstable: if Logdec < 0

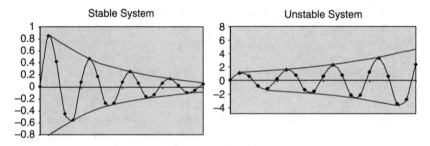

FIGURE 5.3   Representations of stable and unstable rotor systems (Ref. 3).

At 3.2 mil, the whirl orbit would be 3.2/8.0 = 40 percent of the clearance—close to the 50 percent limit and certainly cause for concern.

As compressor users we must try to visualize rotor bow at mid-span. That is why many multiimpeller (multistage) compressors have a stainless steel labyrinth between the last wheel in a given compression stage (or compression section). At, say, 6000 RPM, a severely bowed shaft can contact a portion of the labyrinth 100 times in a single second. The shaft will heat up instantaneously, bow some more, and cause a serious wreck. To avoid this heat-up eventuality, recall that we advocated use of frangible labyrinth materials in an earlier chapter.

Examining What Can Be Done about Instability

Process operations personnel may have ways of juggling their variables (MW, d_{inlet}, d_{exit}, HP, and RPM) in efforts to reduce the magic number Aero C-C. Competent machinery engineers may be able to introduce a little more damping (sponginess, vibration soak-up capability) by changing the lube oil pressure and viscosity (a temperature-dependent property), and also by changing the bearing preloading or orientation of the shaft relative to the bearing surface. This latter fix will require lifting the back end of the compressor by shimming, or thermally increasing the outboard support legs as much as 40 to 50 mil (1–1.25 mm).

But that's of only limited value at times and can gain a little bit of maneuvering space at best. A permanent solution will require changing to a bearing type which provides greater damping. Reliability-focused users often purchase a set of these bearings—"just in case."

Thrust and Radial Tilt Pad Bearings

A typical thrust bearing is shown in Fig. 5.4. The entire assembly consists of a number of tilting pads (six, in this illustration), which are allowed to articulate and operate in conjunction with a thrust-equalizing disc secured to the shaft. The thrust disc shown on the right of Fig. 5.4 is surrounded by a two-piece labyrinth seal.

Thrust bearings are used to locate the rotor axially and at the same time absorb any axial rotor thrust. Thrust bearing configurations include flat land, tapered land, and tilting pad models. Their respective load-carrying capacities range from 50 psi to 250 psi. However, the most popular configurations comprise self-equalizing, leading-edge spray-lubricated tilting pads with permissible specific loads ranging from 200 to 400 psi (Kingsbury-type LEG). Bearing material options include tin/lead base babbitt and various copper-bearing alloys (bronzes) to suit specific applications. In each case, the compressor designer desires not to exceed 50 percent of the bearing manufacturer's allowable load rating. Bearing designs also call for suitable conservatism whenever the possibility exists of additional thrust being transmitted from or through multiple casings.

Figure 5.4   A six-pad thrust bearing assembly (Ref. 4).

Thrust bearings must have the correct axial clearance—typically 0.008 to 0.010 in—to perform properly. During maintenance-related shutdowns and before further dismantling, the exact rotor float from its mid position toward the active and nonactive sides of the thrust bearing should be checked. When the bearing is removed, the pads should be examined for wear and other surface damage. Also, the rotor free-float should again be checked before the compressor is fully assembled at site. Free-float shows the feasible axial impeller movement within the casing before the thrust bearing is being installed. Final readings are taken and recorded after installation of the thrust pads. In essence, the correct positioning of the rotor in its casing and observing thrust collar locations relative to the active and inactive thrust bearing pads require considerable care and patience. Axial position probes (proximity probes) are used for on-stream monitoring of rotor location.

Many modern centrifugal compressors also use tilt-pad bearings of the type shown in Fig. 5.5 in the radial shaft support location. The number of pads ranges from three to seven and, although the pads are equidistant from each other, they can be oriented for load-on-pad or load-between-pad.

A rotor response plot (Fig. 5.6) simulates rotor unbalance and determines rotor speeds where vibration amplitudes would be excessive. Again, the number of pads and load orientation (load-on-pad vs. load-between-pads) influences the damping qualities and determines the range of speeds where a particular compressor rotor will be susceptible to high vibration.

The compressor manufacturer makes a thorough analysis of the rotor behavior under slight unbalanced conditions. The results of such an analysis are shown in Fig. 5.6, an example case for a six-stage rotor. Here, the resulting vibration would be at its most severe,

FIGURE 5.5   A tilt-pad style radial bearing (Ref. 4).

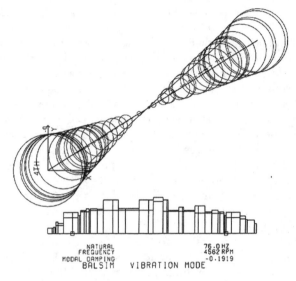

NATURAL
FREQUENCY
MODAL DAMPING
BALSIM   VIBRATION MODE

76.0 HZ
4562 RPM
-0.1919

Result of critical speed analysis

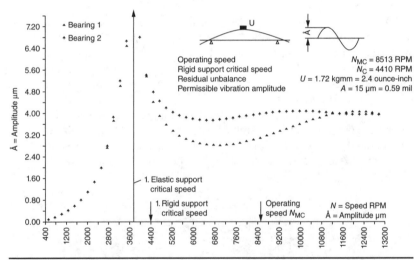

FIGURE **5.6**   Critical speed analysis or unbalanced rotor response plot (Ref. 5).

around 3600 RPM. At the projected operating speed of approximately 8400 RPM of this rotor, the vibration amplitude would be no more than 4 μm—much less than the allowable 15 μm.

Before we leave the subject it is worth mentioning that transient local flow instabilities are another, fortunately quite rare, deviation from smooth operation. Sometimes, a flawed control valve upstream of the compressor inlet is at fault. At other times, a surge or stalled flow condition may occur unexpectedly. Again, these conditions are rare, but we need to know about them. Never discourage expert analysis.

Bearing Clearances

Bearing clearances will affect heat generation and support stiffness. A typical diametral clearance can be calculated from the expression

$$C = (0.001 \times \text{shaft diameter}) + 0.002 \text{ in}$$

As an example, a compressor with a journal diameter of 8.000 in would use bearings with bore diameters of nominally $(0.001 \times 8) + 0.002 = 0.010$ in. One might use a tolerance of $+/-10$ percent and allow bearing bores ranging from 0.009 to 0.011 in.

Centrifugal compressors are generally designed to operate above their first critical speeds. Critical speed is somewhat analogous to the resonant frequency at which a rotor assembly would vibrate if struck by a hammer. Critical speed analyses and rotor response plots must be provided by the vendor and must be reviewed by the buyer's engineer. If the owner-purchaser and vendor-manufacturer disagree on calculated results, a third-party analysis may be required to resolve discrepancies and reconcile the different results.

Although both bearing header and bearing metal temperatures are typically monitored, bearing metal temperatures are the more important of the two. Alarm settings of 110°C to 120°C are generally used for bearing temperatures while high header temperature (bulk oil) alarms are often set at 80°C. Centrifugal compressors should be shut down when the actual bearing temperature reaches 135°C and the cause or reason for this high bearing temperature should be examined.

Tin-based babbitt is preferred over lead babbitt due to its superior corrosion resistance and better bonding with steel backing. The tin-based material is, however, not as "forgiving" of dirt inclusions in the lubricating oil as is the lead-based version of the alloy.

Vibration and Acceptable Limits

Because vibration amplitude is a measure of the severity of the trouble in a machine, an obvious question will be: How much vibration is too much? To answer this question, it is important to bear in mind that the objective is to use vibration checks to detect trouble in its early stages and schedule an appropriate correction procedure. The real goal is to get a fair warning of impending trouble, not to determine how much vibration a machine can tolerate before it will fail.

There are no absolute values or vibration limit which, if exceeded, will result in immediate machinery failure. The events surrounding the development of a mechanical failure are too complex to set any reliable limits. On the other hand, you must have some general indication of machinery condition that can be evaluated on the basis of vibration amplitude. This is possible through the use of general guidelines that have been developed by experience over many years.

The guidelines shown in Fig. 5.7 generally refer to centrifugal compressors that have been purchased and installed in accordance with best practices. Obviously, operation in the field labeled "good" should give you comfort, but what about operating in the "marginal" or "unsafe" ranges? Here's our recommendation. As you move past the mid-range of "marginal," consider procurement and implementation of permanently mounted monitoring instruments. Depending on compressor size, speed, internal geometry, and general accessibility, this instrumentation might include radial and axial casing vibration transducers in addition to the existing shaft-observing eddy current probes. Also, it would be advantageous for the instrumentation package to include thrust bearing temperature thermocouples and shaft position-monitoring probes. The various probes or monitoring instruments

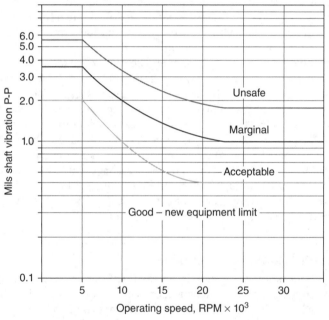

Notes:
1. Operation in the "unsafe" region may lead to near-term failure of the machinery.
2. When operating in the "marginal" region it is advisable to implement continuous monitoring and to make plans for early problem correction.
3. Periodic monitoring is recommended when operating in the "acceptable" range. Observe trends for amplitude increases at relevant frequencies.

FIGURE 5.7 Compressor shaft vibration (1 mil = 0.001 in of radial motion), direct-measured by an eddy current probe mounted in the bearing cap (Ref. 6).

would be connected to logic circuitry or trip logic modules configured and wired to trip the machine whenever two out of three maximum allowable values would be exceeded. The maximum allowable value of a temperature thermocouple sensing true babbitt temperature might be 260°F (126.7°C); the maximum allowable shaft vibration amplitude from one or more of the various eddy current transducers would be as shown by the "unsafe" line in our diagram.

The limits shown in the figure are based on the long-term experience of a practicing engineer.[6] Note that these vibration limits refer to probe installations close to and supported by the bearing housing. Assume that the main vibration levels above 20,000 RPM reflect the experience of high-speed air compressors (up to 50,000 RPM). Readings must be taken on machined surfaces, with runout less than 0.5 mil (0.0005 in) up to 12,000 RPM, and less than 0.25 mil above 12,000 RPM.

Judgment must be used, especially when experiencing frequencies in multiples of operating RPM on machines with standard bearing loads. Such machines cannot operate at the indicated limits for frequencies higher than 1 × RPM. In such cases, enter onto the graph the predominant frequency of vibration instead of the operating speed.

Anyway, when analyzing machinery vibration to pinpoint a particular problem, it is essential to know the vibration frequency. Knowing the frequency helps identify the problem and points to the part that has changed. Forces causing vibration are generated by the rotating motion of machine parts. Because these forces change in direction and amplitude according to the rotational speeds of the various parts, it follows that many vibration problems will have frequencies that are closely related to rotational speeds. We can often spot a defective part by noting its vibration frequency and associating that frequency with the rotational speed of the various machine parts. Alternatively, we can attempt to determine acceptable vibration levels as a function of speed harmonics.

The speed harmonics table, Table 5.1, relates to vibration measured on bearing housings. A velocity transducer would feed the resulting signal into a data collector/analyzer with filtering capability. Such a data acquisition package would be capable of displaying equipment vibration velocity (inches or millimeters per second) at running frequency and its respective multiples.

The harmonics table for various rotating machines shows two rows labeled "A" and "B." Experience shows "A" values to represent maximum filtered velocities on machines clearly considered smooth running. Similarly, "B" values represent the highest levels of short-term operability reported by a fair sampling of the machines listed. It should be noted, however, that not every machine may necessarily still be safe to operate at these levels. Similarly, machines with sturdy, overdesigned components may survive longer than expected.

		Speed Harmonics									
Machine Type:		**1**	**2**	**3**	**4**	**VP1**	**VP2**	**GMF1**	**GMF2**	**BP1**	**BP2**
Blowers	A	0.05	0.02	0.01	0.01	0.04	0.01				
(6000 RPM max.)	B	0.5	0.4	0.25	0.25	0.1	0.05				
Horizontal centrifugal	A	0.05	0.02	0.01	0.01	0.04	0.01				
compressors	B	0.25	0.2	0.15	0.15	0.1	0.05				
Barrel compressor	A	0.03	0.01	0.005	0.005	0.03	0.005				
	B	0.15	0.1	0.1	0.1	0.05	0.025				
Gears: parallel shaft	A	0.1	0.05	0.02	0.02			0.05	0.02		
(Gen. purp.)	B	0.5	0.4	0.25	0.25		0.15	0.1	0.05		
(Spec. purp.)	B	0.25	0.2	0.15	0.15		0.15	0.1	0.05		
Gears: Epicyclic	A	0.05	0.02	0.02	0.02			0.05	0.01		
	B	0.15	0.1	0.1	0.1			0.1	0.05		
Steam turbines	A	0.1	0.02	0.02	0.02					0.05	0.01
(Gen. purp.)	B	0.5	0.4	0.25	0.25					0.1	0.05
(Spec. purp.)	B	0.3	0.25	0.15	0.15					0.1	0.05
Turbines/axial	A	0.2	0.02	0.01	0.01					0.05	
compressor	B	0.5	0.4	0.25	0.25					0.1	0.05
Pumps	A	0.1	0.05	0.01	0.01						
(Outb. bearing)	B	0.25	0.2	0.15	0.15	0.1	0.05				
(Overhung)	B	0.5	0.4	0.25	0.25	0.1	0.05				
Motors	A	0.1	0.1	0.05	0.05						
	B	0.25	0.2	0.15	0.15						
Screw	A	0.1	0.01	LP1 0.1	LP2 0.1	LP3 0.1	LP4 0.05				
compressor	B	0.25	0.2	0.2	0.2	0.2	0.2				

Abbreviations: VP, vane pass; GMF, gear mesh frequency; BP, blade pass; LP, lobe pass.
A: Highest noted on smooth machine.
B: Maximum level of operability (repair as soon as possible).

TABLE 5.1 Compressor Vibration Velocity, Peak Amplitude Readings Measured on Bearing Housing (inch-per-second) per Ref. 6.

A reasonable approximation for maximum levels of operability with unfiltered vibration velocities could be calculated by taking the square root of the sum of the squares. For blowers, this maximum unfiltered vibration velocity limit would be 0.74 in/s.

What We Have Learned

Rotordynamic studies (unbalance response analyses) are needed to define rotor speeds that must be avoided.

Compressor bearing configurations and bore clearances affect rotor stability. Vibration amplitude and frequency monitoring can track rotor behavior and are important.

High-frequency vibrations in high-pressure compressors may be aerodynamic in origin. Anti-swirl vanes or de-swirling labyrinths may need to be installed in such machines.

Bearing temperature tracking can give early warning of discrepancies that must be addressed. Whenever sensors embedded in the babbitt reach 135°C, a precautionary shutdown should be effected.

Experience-based vibration guidelines may not be totally accurate but they are infinitely more precise than mere guesses.

References

1. Salomone Turbo Engineering Inc., Consulting Memo to HPB, Houston, TX, 1990.
2. Demag-DeLaval, Commercial Literature, Duisburg, Germany, 1990.
3. Engineering Dynamics Inc., Marketing Brochures, San Antonio, TX, 1990–2000.
4. Hitachi Ltd., Commercial Bulletin, Tokyo, Japan, 1998.
5. Borsig GmbH, Commercial Literature, Berlin, Germany, 1987.
6. Zierau, S., "Machinery vibration spectrum analysis: a blend of problem solving and advancement of diagnostic know-how," Petrotech Maintenance Symposium, Amsterdam, Netherlands, April 1976.

CHAPTER 6

Lube and Seal Oil Systems

Compressor auxiliaries are responsible for more downtime events than actual compressor components. To be considered reliable, they deserve close scrutiny and must often be upgraded from the traditional vendor's standard configuration. Compliance with API-614 is helpful, but it must be kept in mind that the various API standards are intended to explain minimum requirements. Minimum requirements can differ from best available technology.

As of 2011, only a relatively small percentage of the many thousands of centrifugal compressors operating in modern industry were equipped with magnetically suspended or gas-lubricated bearings. The overwhelming majority of compressors use oil lubrication for the bearings that either support the compressor shaft (radial bearings) or limit shaft axial movement (thrust bearings). This chapter deals with these systems.

Similar comments pertain to compressor seals. Seals are needed to prevent migration from the pressurized compressor interior volume (the compression space) toward the bearings. These seals come in a variety of configurations (see Chapter 4) and the majority requires oil as a coolant and lubricant. The auxiliary systems that feed oil to bearings and seals are often combined, in which case they are aptly called lube and seal oil systems. Separate systems are more common and will be required if seal oil is contaminated by entrained "sour" gases, such as H_2S. A plain lube oil system is represented in the simplified schematic of Fig. 6.1. A few of the most common instruments are also shown on this illustration.

Reservoirs must include valve and space provisions for temporarily or permanently connecting oil purifiers to the low-point drain. In addition to removing water contamination, modern oil purifiers will also remove undesirable gases from the seal oil. Additional comments are found later in this chapter; don't overlook them.

If both drivers are electric motors, different feeder connections are recommended by API-614. It should be of interest that Note 1 (in Fig. 6.1) also alerts purchasers to locate the suction piping away from reservoir low points where dirt might easily accumulate.

56

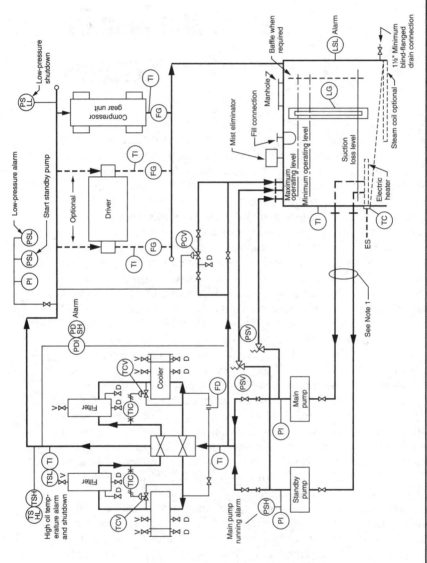

Figure 6.1 Simplified, but typical, compressor lube oil system. Multiunit systems require provisions to separate (to valve-off) one system from another. In combined lube-and-seal oil systems with turbine drivers, the compressor's outer seal oil drain must be separate from the lube oil drain.

AS	Air supply	LSH	Level switch high
FI	Flow indicator	LSHH	Level switch high high
FO	Flow orifice	PDCW	Pressure differential control
FSH	Flow switch high		valve
FSHH	Flow switch high high	PDI	Pressure differential indicator
FY	Solenoid valve	PDIC	Pressure differential indicator
HCW	Hand control valve		controller
LCV	Level control valve	PDT	Pressure differential transmitter
LG	Level glass	PDSH	Pressure differential switch high
LSL	Level switch low	PDSL	Pressure differential switch low

TABLE 6.1   Typical Instrumentation Found on Lube and Seal Oil Systems

A reliability-focused user will take a very active part in the selection and design process for these compressor support systems. An infinite number of component combinations are possible and user preferences are to be discussed and agreed upon at the selection stage. Guidance can be found in the various API specification documents; however, the instrument nomenclature chosen by vendors and manufacturers often differs. Table 6.1 is one of many hundreds of feasible listings of instruments typically found on lube and seal oil systems. The owner-purchaser's engineer must understand the purpose and functionality of each of these elements.

Layout Guidance

All systems must be properly laid out and supply piping sized for maximum velocities not in excess of 7 fps (~2 m/s). Stainless steel is used for all piping, both upstream and downstream of filters. Stainless steel is also needed for vessels, housings, tanks, and their respective tops. Only certain valves and a few instruments are (possibly) exempt from this requirement. With high reliability the first and foremost goal, all supervisory and control instrumentation elements are likely to include stainless steels.

Cost-cutting has made inroads here, although some "savings" are false economy that will often cost dearly. To avoid unavailability, here are some of the key areas that should not be overlooked:

- Access to major hardware and instruments should be easy.
- Filter housings must be vented to a safe location. After replacing a filter, air must be vented so as to make this standby filter housing ready for operation. Venting back to the oil reservoir is allowed.

• With the possible exception of valves, all oil-wetted parts of the lube oil system (but not the pumps) should be made of stainless steel. The top lid of the oil reservoir must be made of stainless steel also, because moisture condensation can accumulate on this cover.

• The switch-over valve directing oil through either the "A" or the "B" filter-cooler set must incorporate provisions to lift its plug off the valve seal before the plug can be rotated in the desired direction.

• If the top lid is made of plain steel, the resulting rust formation (on the inside) will take its toll as reduced equipment reliability or require increased preventive maintenance. It should be noted that a nitrogen "blanket" to fill the space between liquid oil and top lid will not be a fully effective method of preventing rust on plain steel top lids.

• The top lid is slightly inclined to allow rainwater and spilled oil to drain. Pipe connections and access ports ("manways") are flanged with top openings raised at least 1 in above the reservoir top and no tapped holes are allowed anywhere on the reservoir.

• All fill openings must be provided with removable strainers.

• Integral internal relief valves are permitted on rotary positive displacement pumps. However, only external relief valves are permitted on pressure vessels.

A small-to-mid sized lube skid is shown in Fig. 6.2. The photo depicts two vertically arranged filters and a manual lever-operated switching valve located near the base and between the two vertical filter housings. Only one heat exchanger was apparently selected by the customer; it can be seen on the right side of the skid. Oil returned from the lubricant user—perhaps a small compressor—reenters the stainless steel reservoir near the upper left corner of the photo. Although no lube oil pumps are shown in Fig. 6.2, the skid designer will undoubtedly have sized at least two, and sometimes three lube pumps for oil flows that include unusual upset conditions. When both pumps are motor driven, different feeders or a DC supply source are generally specified. The direct-current source must supply power for as long as it takes an operator to secure the main compressor and manipulate all associated valves.

Again, note how the principal components are readily accessible. Suction pipes must be arranged to provide positive suction head for the various feed pumps, with the suction line sloped down from the reservoir to horizontal pumps, or with at least two (and sometimes three) vertical-style feed pumps actually immersed in the oil residing in the reservoir. Our recommendation to install horizontal piping

Figure 6.2 In this accessible skid-mounted lube oil system, the filters are in the right foreground; a single cooler (heat exchanger) is horizontally arranged on the right (Ref. 1).

with a slope is sometimes contested by pump manufacturers. However, sloping will allow gas to be vented back to the reservoir; it should be considered mandatory.

To rule out unexpected surprises and the occasional finger pointing, the compressor manufacturer must be directly responsible for the design, although the manufacturer often asks third parties to fabricate and test the entire skid.

Examine What Often Goes Wrong

Reliability-focused users specify lube and seal oil systems that comply with the applicable standards of the American Petroleum Institute (API-614). These standards constitute a detailed and enhanced bill of materials as well as a description of redundancies required to impart years of uninterrupted uptime to such systems. Appropriate instrumentation must be provided and an experienced compressor operator should be involved in selecting these instruments and determining their operator-friendly, optimum, mounting locations. Ease of maintenance and accessibility compete with the desire to keep things compact and a measure of judgment must be exercised by purchaser and vendor.

With few exceptions, systems that do not comply with API Standards will require more frequent maintenance. Regardless of standards applied, the purchaser would be wise to review a number of pertinent details. Here are the ones most often overlooked:

Main versus Standby Pump

Pumps must be centrifugal or rotary-positive displacement. Driving off the main driver or compressor shaft is rarely acceptable, because pump failure would mandate equipment shutdown. If two or three pumps are used, at least one is usually driven by a small steam turbine. Pumps must have carbon steel casings, and cast iron casings are allowed only inside the reservoir. Exposed cast iron pumps would be prone to crack when involved, directly or indirectly, in a fire event.

A decision must be made as to which pump is normally on standby (although the turbine-driven pump is usually selected for standby duty). Still, someone must define how quickly the turbine comes up to speed (and reestablishes the required oil pressure), and what the electrical classification should be for motor drivers. Suitable electronic governors should be selected for the small steam turbine. If the steam turbine driven pump is in standby mode, it might have to be kept warm and "slow-rolled." It should have a return line with restriction orifice back to suction, and dewatering of piping and steam turbine casing must be accomplished by using the right steam trap types and models.

Standby equipment deserves more attention than it usually seems to receive. Pumps and their respective driver shafts must be relatively easy to align.[2] Couplings should be specified with a service factor of 2 or more, and these should be relatively maintenance free. Drivers should be specified with load factors or service ratings that correspond with best practices.

The start switch or actuator component for the auxiliary pump must have a manual reset provision. A steam condensate exhaust hood will be needed for steam exhaust lines to atmosphere. Without it, operators risk being showered with scalding water whenever the auxiliary steam turbine-driven pump set kicks in.

Slow-Roll Precautions

On some steam turbine models, slow rolling below a speed of approximately 150 rpm will not allow an oil film to be established between journal and bearing bore. Also, consideration must be given to an emergency source of oil to be fed to the turbocompressor train in an occasional power failure event. If there is even a remote possibility of neither oil pump being available, an overhead rundown tank should be provided to gravity-feed the turbomachinery bearings. A pressurized overhead tank is shown in Fig. 6.3, but nonpressurized (atmospheric pressure) tanks are quite often used as well. An atmospheric

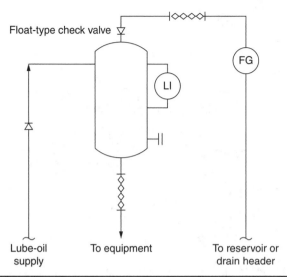

FIGURE 6.3   Pressurized overhead rundown tank for centrifugal compressors (Ref. 3).

breather valve or vent must be used with nonpressurized models and the user-purchaser must pay attention to issues of airborne dirt and birds trying to build nests in or near such vents. A drilled check valve is then used between the lube supply header and atmospheric pressure overhead rundown tank. Regardless of the type of rundown tank selected, elevations must be such that the static head is less than the equipment lube-oil trip pressure. An API standard (API-614) gives guidance on these and several other important matters dealing with lubrication, shaft sealing, and control oil systems for special purpose applications.

The anticipated time it typically takes for the machine to coast to a stop is 8 minutes, with 15 minutes a more conservative limit. This rundown tank should be vented and the vent must be oriented and configured to prevent entry of birds and debris. Ask if the overhead rundown tank must be heated or insulated for operation in cold weather. Are suitable autostart facilities provided? Double-check to verify proper dewatering facilities provided at all points of the steam piping and at the turbine casing. Ask also if these facilities are reliable or were simply purchased from the lowest bidder without further thought given to maintenance requirements and energy efficiency.

In installations with two electric motor-driven pumps, the power should come from different feeders or substations. Temporary power dips during pump switch-over are bridged by using a hydraulic accumulator in the lube supply line. The bladder of the

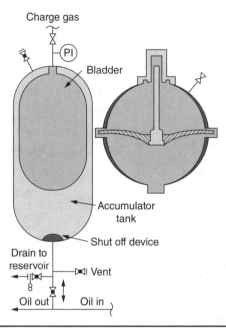

Figure 6.4   Bladder-type accumulator (left) and a rod-equipped "surveillable" diaphragm-type accumulator (right) (Ref. 5).

accumulator is usually filled with nitrogen and the configurations and functionalities of such accumulators are well known. Yet, although relatively widely used, typical bladder-type accumulators (Fig. 6.4, left image) risk premature failure from the rubbing action of the neoprene or Buna rubber bladder against the accumulator walls. This failure risk is further amplified when dirt particles are carried in the oil.[4] Diaphragm-style accumulators (Fig. 6.4, right image) were used in reliability-focused user companies after 1975 to facilitate condition monitoring and to avoid such rubbing-induced failures. Note (right image) that a standard diaphragm-type accumulator is fitted with a vertical indicator rod and a transparent dome at the top.

Reliability-focused plants modify the standard diaphragm accumulator of Fig. 6.5 by removing seal ring and screw plug and tightly fitting a tall transparent high-strength plastic dome at the top of the accumulator. A tapped hole is machined into the center of the shut-off button and a long "gauge rod" threaded into this tapped hole. The gauge rod extends through the opening created by removing the screw plug. The tip of the gauge rod is seen by the operator making their surveillance rounds. The integrity of the flexible diaphragm and its properly proportioned nitrogen versus oil-fill volumes are visually ascertained as is shown in Fig. 6.6, which shows a large field

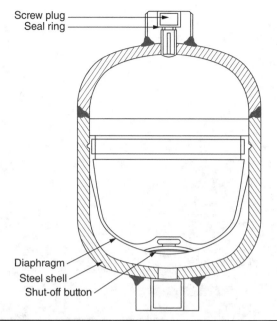

Screw plug
Seal ring

Diaphragm
Steel shell
Shut-off button

Figure 6.5 Diaphragm-type accumulator (Ref. 6).

installation. Wire mesh screens are installed to guard against a careless overhead hook or a maintenance tool accidentally hitting the polycarbonate sight glass dome.

If bladder-type accumulators are deemed acceptable, be sure they have a 10-second or greater capacity and are equipped with fill valves and isolation valves that permit monitoring of bladder condition. Bladderless accumulators will require high-level alarm, low-level alarm, and low-level cut-off provisions.

System Reservoirs

An armored sight glass must be supplied for the reservoir. Because the reservoir is to be made of stainless steel, its interior is not to be coated or painted.

Minimum standard practice calls for oil reservoirs to be sized for at least 2.6 minutes of maximum flow. In other words, a lube oil system with pumps supplying 100 gpm would be sized for an operating volume of 260 gallons (1000 L) or more. A more conservative high-reliability practice defines the system operating range as 2.6 times gpm, to which is added either 40 gallons or 1 week's oil leakage rate, whichever is more. Other rules-of-thumb are noteworthy; one of these calls for a free oil surface in the reservoir of at least 0.25 ft × 2/gpm so as to promote air disengagement from the oil.

FIGURE 6.6 Diaphragm accumulators installed at a best-of-class facility.

Oil reservoirs are typically rectangular and are provided with a sloped bottom, sometimes called a "false bottom." The volume below the sloped false bottom is filled with a heat transfer fluid for pre-startup heating or for maintaining a controlled temperature. The volume above the false bottom is, of course, the actual working volume of the oil reservoir. Convention calls for a reservoir vent to be one pipe size larger than the sum of the areas of all seal drains.

In installations where steam is available, a thermal fluid with high temperature capability and low volatility should fill the space below the sloped bottom. If no steam is available, electric heaters sized not to exceed 15 W/square in (the "watt density") can be used to heat the thermal fluid. Electric temperature control switches should be provided if electric heat is selected. A high-capacity vent is needed to accommodate thermal expansion of the heat transfer fluid below the sloped bottom of an oil reservoir. A side-mounted gauge glass or dipstick is required to verify or monitor the height of thermal fluid under the false reservoir bottom. If a steam coil is used for heating, there should be suitable steam traps.

Heating Requirements

In some climates, heating will be needed only at startup or in low-temperature ambient conditions. Heaters are generally sized to effect heating from lowest average ambient to a minimum allowable oil temperature (73°F/20°C, for the very typical ISO VG 32 lubricant) in 4 hours. It is possible to preheat the lubricant by simply admitting steam into the water upstream of coolers. However, temperature indicators will have to be installed and a responsible individual will have to be put in charge of this emergency heating operation. For cold temperature regions or in situations where large ambient temperature swings are common, the reservoir may require external insulation. Such insulation then has the associated benefit of reducing condensation of water vapors in the reservoir.

The return oil from the turbocompressor may be at sufficiently elevated temperature to flow freely without further addition of heat. Valves are required at the low points of the reservoir working volume and at the lowest possible point of the heating space below the sloped reservoir bottom. The drain valve at the low point of the working volume serves also as a connection for an on-stream lube oil purifier. Such purifiers are normally sized to handle the entire system working volume in 24 hours. They must be provided with a piping leg that prevents reservoir emptying. Some will also require a condensate removal line.

All reservoirs must be fitted with internal baffles or stilling tubes that allow for contaminants to settle out. Oil returning from the turbocompressor bearings or bypassed from the pressurizing pumps is not allowed to fall into the reservoir since that would risk static electricity build-up. The vents from filter housings and other points in the installation should lead back into the reservoir.

Filters and Coolers

Suitable instrumentation is needed on filters and coolers. The layout must permit the system to operate while maintenance personnel are safely performing routine service on nonoperating redundant elements. Except for the transfer valve (main switching valve) and the structural parts of the mounting skid, stainless steel is the required material of construction in reliability-focused plants. Block valves and check valves are needed and the user-purchaser must devote time and effort to review the piping and instrumentation diagram (P&ID) for functional completeness.

Kickback valves that route excess oil back to the reservoir must be located upstream of the filters and coolers. They should be sized to pass the excess capacity of one pump plus the full capacity of the standby pump. Dual valves may be needed to obtain proper valve coefficients in certain seal systems.

It is usually considered a good move to involve one's operators before specifying and purchasing filters and coolers for an existing facility. Blotter paper-style filter cartridges are not acceptable. Allow the operators to ask if they are satisfied with the instrumentation package shown on the schematics, or on a mockup of the system. Obtaining buy-in from staff at this stage is of great future value.

Normally, coolers are bought in compliance with TEMA Class C shell requirements and have removable bundles. Experienced users will not permit tubes with less than 5/8 in OD 18 BWG, but will allow double pipe fin-tube exchangers for small systems. It is usually best to check prior user experience and ask questions. Unlike pumps, coolers are pressure vessels that must be designed and manufactured in accordance with applicable codes. Water should be on the tube side, oil on the shell side. The oil pressure must exceed cooling water pressure to prevent, or at least reduce, leakage of water into the oil system in the event of tube failure. The oil side design pressure should be equal to, or greater than, the pump relief valve setting with PD pumps and shutoff pressure with centrifugal pumps. Material selection guidelines are given in Table 6.2.

Each cooler must have an oil fill line, a drain, and a high point vent—all suitably valved and generously sloped. Cooling water flow enters at the bottom and exits at the top. Drain piping is typically sized for a maximum velocity of 1 fps (~0.3 m/s).

High-pressure gas piping should be seal welded and all piping should be configured to allow for its thermal expansion. Remember that the piping may have to be removed for cleaning prior to compressor commissioning. There need to be flanges and special locations (such as near bearings and seals) for the insertion of temporary strainers. Flexible expansion joints are not allowed in the piping because of the danger of fatigue failure. Flexible joints and hoses are also disallowed because they tend to be the first stationary elements to fail in a fire.

To facilitate oil drainage back to reservoirs in typical gravity systems, each compressor bearing housing typically requires a 1-in minimum vent. Gear boxes and couplings are generally equipped with 2-in vents. Coupling guards may require special air exchange provisions to prevent trapped air from overheating the coupling components.

Shell	Channels and Covers		Tube Sheets		Tubes	
	Material	Specification	Material	Specification	Material	Specification
Carbon steel	Acid resisting bronze or aluminium bronze	ASTM B143 Alloy 2A ASTM B169 Alloy 614	Naval brass	ASTM ASTM B111 Alloy 464	Inhibited Admiralty	ASTM B111 Types 443, 444, or 445

TABLE 6.2 Material Selection for Heat Exchangers Used on Compressor Lube Skids

Centrifugal Compressor Lube/Seal Reservoir Explosion Hazards

The static electric charge generation mechanism was investigated by prominent users in the mid 1970s. Static charge buildup in filters was determined to be the cause; it was confirmed by careful measurements. Systems that had experienced explosions were equipped with pressure-controlled recycle lines downstream of the seal oil (or lube/ seal oil) filters. In obvious contrast, systems with recycle lines originating upstream of filters and with line lengths that allowed relaxation of charges remained trouble free.

Safe designs allow 30 or more seconds for the oil to travel from filter outlet to reservoir inlet. Because undesirable agitation of the oil surface must be avoided, the return line should enter the reservoir below the oil level. Pressurized return lines should not be vented inside the reservoir.

Seal oil system gas reference lines should be provided with a drilled check valve to prevent disruption of overhead accumulator level control during compressor surge. There is also a need for provisions that allow introduction of a simulated gas signal (sometimes called a "false buffer gas") during startup when running a compressor on air, or with a suction pressure below design. These provisions may require control systems that can fully accommodate prevailing running-in conditions.

What We Have Learned

Lube and seal oil systems must be carefully and conservatively designed. A review of their adequacy must start at the proposal stage. The owner-purchaser must thoroughly understand each design element. Each pipe or control line should be traced back to its origin; its design intent must be understood, and all of the owner-purchaser's questions must be answered by the vendor-manufacturer.

Reliability-focused owner-purchasers go beyond the minimum requirements of API-614 in their efforts to impart the ultimate in maintainability and ease of surveillance to these very important systems.

Special diaphragm-style accumulators are one of many examples where reliability-focused thinking is translated into component selection. They represent best-available technology and have been used by best-of-class companies for many decades.

References

1. Lubrication Systems Inc., Division of Colfax Industries, Houston, TX.
2. Bloch, Heinz P., *Pump Wisdom: Problem Solving for Operators and Specialists*, John Wiley & Sons, Hoboken, NJ, 2011.

3. American Petroleum Institute, API Standard 614, Alexandria, VA, 1997.
4. D'Innocenzio, Michael, "Oil systems—design for reliability," Proceedings of First TAMU Turbomachinery Symposium, College Station, TX, 1971.
5. Bloch, Heinz P., "Making machinery surveillable," *Hydrocarbon Processing*, July 1993.
6. Doddannavar, Ravi and Andries Barnard, *Practical Hydraulic Systems*," Elsevier Publishing, Burlington, MA, 2005.

CHAPTER 7

Impellers and Rotors

Construction, Inspection, and Repair of Impellers

A closed impeller with backward-leaning vanes is shown in Fig. 7.1. Such impellers offer a wide operating range and the backward lean offers a combination of process stability and reasonable efficiency. Two versions are widely available, a two-dimensional (2-D) and a three-dimensional (3-D) configuration. Figure 7.2 shows a 3-D impeller that was contour milled from one single piece of metal. This type of contour milling is ranked among the favored fabrication processes and is not to be confused with techniques whereby individual blades are being welded to an impeller hub.[1]

Two-Dimensional versus Three-Dimensional Impeller Blading

The backward lean in a 2-D version has the same curvature throughout the blade width. In the case of a single shaft compressor with multiple stages, the volume at the inlet to the next stage is being reduced and this lowers the stage efficiency. A compromise comes to mind: One could reduce the impeller diameter to gain efficiency. However, reducing the diameter would reduce head produced and more stages would be needed. Now the shaft would have to be made longer, which would affect the dynamic stability or susceptibility to undesirable vibration behavior of the compressor rotor.

All of these factors must be given consideration for good compressor design practices. The same machine will not incorporate significantly different impeller diameters on the same rotor. In any event, three-dimensional (3-D) impellers with contoured ("twisted") blades seem to be better adapted to varying flow conditions. For technical reasons, the axial width of a 3-D stage exceeds that of the equivalent 2-D impeller. Since the axial dimension is wider compared to the 2-D version, the number of impellers ("wheels") that can be installed is restricted due to stability considerations. Moreover,

FIGURE 7.1   Typical closed compressor impeller. (*Source:* Mitsubishi Heavy Industries.)

speeds are generally governed by the mechanical tip speed limitations of the impeller material selected by the designer.

The operating range of compressor impellers can also be improved by using a measure of backward lean for the radial portion. Impeller performance and surge point location are related; the surge point determines the useful operating range and head capability. Comparing

FIGURE 7.2   Open impeller with twisted (three-dimensional) blades milled from a solid block of metal. (*Source:* Hitachi Industries.)

three-dimensional impellers with backward-lean radial inlet impellers is of interest. The more highly contoured three-dimensional geometries often feature improved operating and flow range.

Careful Design Needed to Avoid Failures

There are, in certain instances, compelling reasons for using three-dimensional impellers. Vendors have two manufacturing options: that of producing individual blades and to then weld these on an impeller hub, and that of contour machining the entire wheel (the terms "wheel" and impeller are used interchangeably).

In either case, the designer must realize the importance of industry's decade-old practice of not allowing blade passing frequencies (BPFs) to coincide with impeller's natural frequencies. For risk reduction, it will be necessary to validate the vendor's analytical "model" against actual field experience. Aerodynamic excitation could originate in many ways, including in the form of gas returning from downstream impeller(s) to the first-stage impeller inlet, via the balance line. Consideration should also be given to nonuniform and nonrepeatable impeller blade and blade weld geometry and their potential effect on component strength, stiffness, and frequency response.

The inspection and repair decisions for centrifugal compressors may encompass a number of areas. A competent inspector is essential and it should not be assumed that the vendor-manufacturer has a large inspection workforce in place. Inspection coverage must be pre-arranged by the user-purchaser so as to reduce unforeseen startup delays and avoidable failures. Always recall that the compressor owner or its representative must reach a clear understanding with the manufacturer or the compressor rebuilding facility as to the exact deliverables. The achievement of predefined fits and dimensions must be part of this understanding.

While the compressor owner-purchaser has no intention of acting as the compressor designer, many items covered in this chapter are on impellers and rotors. Understanding a few of these items should enable the owner (or owner's representative) to ask relevant questions. The manufacturer's answers will uncover deviations or manufacturing compromises that need to be corrected in the interest of failure risk reduction. An owner's engineer is designated to be the focal point individual whose ultimate task is to ensure that compressors are reliable. While this owner's engineer will probably have many roles to fulfill, space limitations mandate that we select a role that is largely representative of the many other roles and responsibilities assigned to reliability professionals. So, we picked an inspection overview and deliberately chose to explain impeller-related repair and inspection steps.

Compressor impellers are typically custom-fabricated components. Although standard components (bearings, shaft seals, etc) may not merit the same level of attention or inspection coverage, the owner-purchaser may apply similar thinking to verify the adequacy of those and other components. For failure risk reduction an owner's inspector (or a third-party agency inspector reporting to the owner's machinery engineer) must inspect the following:

- Disassembly (unstacking) of the rotors.
- Reconditioning and certification of the shaft.
- In case of upgrading, reconditioning and certification are needed for individual, previously used impellers. Observation of fabrication steps and certification are needed for new impellers.
- Finally, restacking and certification of modified rotors is to be observed and compared with all applicable procedures.

Specific guidelines and areas of concern are included in this segment of our text and should be discussed with the compressor manufacturer or rebuild shop. Then, agreement should be reached on deliverables and written into the contract. Here are some highlights:

Rotor Manufacturing for Centrifugal Compressors

Unstacking the Rotor   An existing compressor rotor such as the one shown in Fig. 7.3 may have to be unstacked (impellers removed) for repair or upgrading. Prior experience shows that careless unstacking procedures cause extreme deformation of impeller bores. Careless procedures (or no procedures) have occasionally been observed even at legacy-brand manufacturers' facilities. That said, a reliability-focused user wishes to avoid repetition of this problem and will insist on having the work performed properly.

1. On old-style centrifugal rotor assemblies, impellers are sometimes secured to the shaft with keys and keyways. However, looseness of either the impeller or the keys is not allowed because it would lead to fretting and unbalance problems.

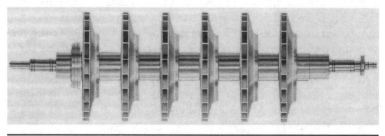

Figure 7.3   A six-stage rotor with closed impellers. The thrust disc is on the right side in this photo; the balance piston is shown on the left.

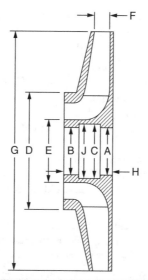

FIGURE 7.4   A typical compressor impeller (also called a "wheel").

2. In Fig. 7.4 the impeller bore region at diameter "A" is called the impeller heel; the bore region at "B" is called the impeller toe. It should be noted that the interference at the toe of many impellers is deliberately made only one-half (50%) of the interference at the heel. Some vendors use these obviously different fits so as to allow a measure of thermal expansion in the direction of the toe while keeping the heel from moving.

3. As we visualize a rotor coming up to speed and going through a resonant speed (sometimes called a rotor critical speed), the shaft will go through a flexing ("bowed") condition. Allowing for a very small amount of sliding movement in the axial direction under the impeller toe will make it less likely for the impeller to clamp itself to the shaft while in a momentary bowed condition.

   There is also a purposely undercut (relieved) region, diameters "J" and "C". Each of these two diameters is related to the keyway dimension in the impeller bore. Also, two keyways are used in some rotor designs.

4. On most modern centrifugal rotors the impeller-to-shaft assembly is made without keys. Instead, there is a uniform impeller-to-shaft shrink fit (0.00075-0.0015 in/in of shaft diameter). Some compressor manufacturers use a uniform (identical) diameter for both toe and heel region. The user-purchaser may inquire about the manufacturer's experience and ask for clarifying answers.

5. In any event, the unstacking procedure will require impeller heating or, in extreme cases, a combination process of heating the impeller and cooling the shaft.

6. The shrink fits are generally calculated to be released when the wheel is heated to 600°F (316°C) maximum. To exceed this figure on materials other than AISI 410 or 4140 could result in metallurgical changes in the impeller. (Note that the impeller is sometimes called "wheel." For the more typically used impellers [made of AISI 4140], one allows a maximum temperature of 1000°F [538°C]. Color-coded crayons which melt at a particular temperature—a popular brand is called "Tempilsticks"—should be used to ensure this temperature is not exceeded.)

7. The rotor is vertically suspended over a sand pit. For unstacking, the lowermost impeller is located at a height of perhaps 2 m (6–7 ft) off the sand surface.

8. The entire diameter of an impeller must be uniformly heated using "rosebud" tips—two or more torches are used at the same time.

9. The important thing to remember when removing impellers is that the heat must be applied quickly to the rim section first. Only after the rim has been heated should heat be applied to the hub section, starting at the outside. A special note deserves to be highlighted: *Never apply heat toward the bore while the remainder of the impeller is cool.*

10. To disassemble rotors, the parts should be carefully marked with a felt-tipped pen or similar indelible marker in the sequence they are taken apart. This will make it possible to later reposition parts in the proper position. A sketch of rotor component position should be made using the thrust collar or shoulder to adjacent impeller hub exit area. Measure and record the distances between all impellers. Each impeller should be stenciled. From the thrust end, the first impeller should be stenciled T-1, the second one T-2, and so on. If working from the coupling end, stencil the first wheel C-1, the second wheel C-2, and so on. This requirement would not be significant in a rerate job where only some of the impellers are being reused. However, there are situations where, after unstacking, the equipment owner might experience an emergency and would ask for the rotor to be quickly restacked and shipped back to the owner's facility. Following the prescribed marking procedure would be of extreme importance if such a need should develop unexpectedly.

11. As mentioned above, a rotor should preferably be suspended vertically above a sandbox to soften the impact of the impeller as it falls from the shaft. Alternatively, the rotors may be suspended over wooden scaffolding or similar rigging as long as the drop distance from impeller edge to the wooden structure does not exceed 2 in (~50 mm).

It may be necessary to tap the heated impeller with a lead hammer in order to get it moving. The weight of the impeller should cause it to move when it is hot enough. Always insist on using a lead hammer; steel hammers are unacceptable for this task. Involving two workers with two hammers applying reasonable blows from diagonally opposite positions makes much sense.

Impeller Inspection and Overspeed Testing   Impeller inspection and overspeed testing are required for new impellers and also for previously used impellers slated for use as reconfigured or upgraded impellers in new rotors.

1. Impeller inspection is divided into two segments, before overspeed and after overspeed tests. Before overspeed testing, the owner's inspector must be present when the compressor manufacturer (or entity responsible for equipment rebuilding) conducts the following operations:

   • Ultrasonic testing of forgings prior to machining

   • Liquid penetrant testing or, alternatively,

   • Magnetic particle examination

   • Measurement and recording of critical dimensions

   If ultrasonic testing has been or will be performed at the foundry, the inspector must review applicable certificates and submit these to the ultimate equipment owner.

2. Liquid penetrant testing or, alternatively, magnetic particle examination is to be performed after each weld operation and after each heat cycle.

3. Impellers must be individually balanced before overspeed testing. Grinding to achieve proper balance should in no case reduce remaining material thickness below the drawing-specified dimension. If necessary, the vendor-manufacturer shall machine an entire quadrant or similar portion of the impeller.

4. Before overspeed testing, the designated inspector must visually examine the cover, disc, and vanes for surface flaws. This should be done at the same time as preliminary liquid penetrant or magnetic particle examinations are underway.

Fabrication Inspection   A number of guidelines cover the fabrication and inspection procedures, and workmanship standards for welded impellers must be as discussed below:

1. Materials
   (a) Impeller materials such as 4140 used in most (but not all) compressor rotors fall within the requirements for Grade B of ASTM A-294, an Ni-Cr-Mo alloy steel. This steel can, theoretically, be heat treated to moderately high yield strengths of 80,000 to 100,000 psi, and ultimate strengths of 110,000 to 130,000 psi. Rotors in H_2S-containing process gas service have yield strength and hardness limitations of 90,000 psi and RC-22, respectively.

   Hardness limits imposed by H_2S service are an indirect limitation on yield strength because of the correlation that exists between tensile strength and hardness (see ASTM A-370). Heat treatments must be controlled so that the above hardness limits are not exceeded on any impellers used in H_2S-containing services. However, the hardness limit may be exceeded in the weld region of impellers whose critical dimensions have been fully established.

   (b) The compressor manufacturer or competent rebuild facility may start with annealed material and heat treat the completed wheel to obtain the desired physical properties. Alternatively, these entities may begin with quenched and tempered material and postweld treat the assembly.

   (c) Determine if the impeller material requires that the parts be preheated and kept heated during welding. Establish if the weldment must receive postweld heat treatment. Failure to keep some materials hot for welding will cause under-bead cracking.

2. Wheel Assembly: Compressor manufacturers, of course, have many wheel configurations, sizes, and wheel designs. The design controls the sequence of assembly, the weld joint configuration, welding process, heat treatment used, etc. The owner's inspector should:

   (a) Determine the manufacturer's methods in building a wheel, including their workmanship standards. If workmanship is not considered acceptable, this must be resolved through discussion and agreement with the manufacturer and the owner's machinery engineer.

   (b) Resolution should start at a pre-inspection meeting. At this meeting, compressor manufacturer and compressor owner-purchaser must reach an understanding of deliverables. It must be kept in mind that if standards need to be improved, the requests must be made in such a manner that extra charges are avoided. For example, there has been difficulty convincing a few manufacturers that they should be concerned about undercutting in welds on impellers. Many competent user-purchasers believe the amount of undercutting permitted should not exceed the following values:

- Maximum depth 0.030 in, up to 1 in long
- Maximum depth 0.010 in, up to 6 in long
- Individual linear indications shall not exceed 3/16-in

(c) Competent user-purchasers also have positions on other weld flaws, including pinholes, cracks, and other defects. Among these are:

- Concavity beyond the drawing-specified crown must be ground down.
- Fillets beyond a specified leg length must be maintained. This is an indirect control on maximum throat thickness.
- The weld bead must be fused at the root and toes. Root pore is not to exceed 1/32-in diameter.
- Occasionally, an excessive gap at the vane-disc or cover interface causes root cracking. If full penetration tee welds are not required at the wheel edge, the cross-section of the weld on the machined edge of the wheel should be visually checked for root cracks and repaired, if any are found.
- The vendor-manufacturer should submit (and commit to using) a procedure that reflects its long-term satisfactory experience.
- Pinholes (piping) should not exceed a maximum diameter of 1/16 in, and not more than one is permitted in each 4 in of weld length. To the extent that deviations are allowed for certain materials, these must be discussed with the owner's machinery engineer.
- Transverse cracks should not be permitted. The owner's engineer may allow other cracks, but these must clearly comply with the manufacturer's long-term experience and must be evident from a written procedure used by the vendor-manufacturer.
- Notches, slag pockets, and arc craters: On unfinished impellers, remove by grinding unless the remaining weld metal is under specified thickness in which case the area should be filled with clean weld metal. On finished

impellers, invoke the compressor manufacturer's experience-based procedure documents.

- Insist on all spatter being removed and do not permit lack of fusion in the transverse direction.

- Require that the compressor manufacturer provide maps showing indications requiring repair on impellers that are to be reused in uprate situations.

- The compressor manufacturer should have experience-based procedures that govern base metal indications and their removal on new and/or used impellers. Be sure that these are followed on your manufacturing or remanufacturing jobs.

(d) Unless approved by the owner's machinery engineer, the inspector cannot incur extra charges to obtain improved workmanship. To the extent necessary, the owner's inspector must verify these indications and request the manufacturer's formal advice regarding cost of required repairs.

3. Impeller Inspection: The owner's inspector should spot-check the weld shop periodically to see that the compressor manufacturer's own procedures are being followed. These checks should be performed at agreed-upon times, if necessary with a manufacturer's escort in proprietary areas of the plant. This type of checking or verification should cover:

(a) Joint preparation. A good many designs call for the vanes to be double fillet welded to the disc and cover, although butt welds and a slot weld have been used. At points of high stress, as on the eye end of vane and possibly at the outer end, complete root penetration may be specified. This requires some type of back-grinding, gouging, etc, after one side of the fillet is made.

(b) Measure the amount of preheat and inter-pass temperature being maintained. Low temperature can cause under-bead cracking.[1]

(c) Check whether correct electrodes are being used and if they are being properly cared for. Wheel welds may be made in one or two passes. AWS Class E 7018 electrodes are commonly used for the root pass of two pass welds or for one pass deposits. This electrode has a low hydrogen coating and good resistance to cracking.

(d) For the second and final layer of weld metal, an E 6027 electrode has typically been used. This electrode produces flat or slightly concave fillets with fine ripples which minimize the amount of cleaning and finishing required. The E 7018 electrode deposit is not quite as good in this respect. Some undercut may be found along the edges where it is difficult to get the electrode in the right position, as in the gas passages. More spatter can also be expected from the 7018 electrode. Note that E 7018 electrodes must be kept in an oven at 225°F (108°C) until actual use.

(e) Both electrodes have lower strength than the parent metal. It is estimated that the weld deposit of an E 7018 wire, through alloy pickup from the base metal and final heat treatment, might end up with a tensile of 75,000 psi. The tensile of the outer layer of weld metal E 6027 might be between 60,000 and 65,000 psi upon completion. This approach has proven satisfactory for compressor wheels. However, some experts believe the strength of the weld deposit should match or slightly exceed the strength of the base metal.

(f) Whenever the finished impeller has a Rockwell C hardness limitation (typically RC 24 as the upper limit) due to H_2S service, the base metal in each wheel should be checked with a portable hardness-monitoring instrument.

(g) Accurate determination of weld metal and HAZ (heat-affected zone) hardness on a finished impeller is most difficult without destroying an impeller. The predefined and specified hardness requirement for the weld can be satisfied by having the compressor or impeller manufacturer make a mock-up joint, using the welding procedure employing the maximum thickness of impeller material to be used, and duplicating the same joint, electrodes, heat treatment, etc, each impeller receives. The mock-up should be cut so the cross-section of the weld is exposed and a Rockwell C hardness traverse can be taken across the face. The traverse should be made parallel to and not more than 2 mm below the surface. If high hardness is verified, the mock-up must be heat treated, resectioned, and rechecked, etc, until satisfactory hardness is obtained. The welding procedure, heat treatment, etc, which produced acceptable hardness levels must be used on the wheels.

(h) This qualification test need not be repeated so long as none of the essential variables are changed. The impeller manufacturer should keep the results of this test on file, for at least 5 years. In other words, if it can be established that the qualification procedure has been followed previously for impellers of identical material, none of the above may have to be invoked.

4. Weld Examination
  (a) Radiography
    (1) Radiography can also be used for checking weld quality in welded impellers, although its use is not widespread due to the type of welding methods used and because of wheel configuration. If radiography is used to inspect welded impellers, the acceptance level for weld flaws must be determined at the preinspection meeting.

    Moreover, it is necessary to agree how often radiography will be taken and what follow-up is required when defective welding is found. The inspector should then obtain the owner's machinery engineer's approval of the applicable manufacturer's standards and procedures. In turn, this engineer may refer the matter to others, but it's his or her ultimate responsibility to seek definition and resolution.

    (2) The quality of the radiographs in terms of density, sensitivity, etc, should correspond to ASME Section VIII, Paragraph UW-51 standards, bearing in mind that weld configuration and impeller construction may prevent strict compliance with code requirements.
  (b) Liquid penetrant
    (1) This technique only discloses flaws open to the surface. The fluorescent penetrants are more sensitive than visible dyes because of the viewing conditions.

    (2) If the vendor opts to use liquid penetrant, his standards should always be reviewed by the inspector. Cracks and crack-like indications are unacceptable. Scattered porosity can be accepted provided there are less than four rounded pores in a line, separated by more than 1/16 in edge-to-edge, axially oriented with respect to the weld. Gross surface porosity density should not exceed that indicated by the medium porosity chart for ½ in thick welds in Appendix IV of Section VIII, Division 1 of the ASME Code. More relaxed standards must be approved by the owner-purchaser's machinery engineer. The weld surface flaw standards in the AWS Structural Welding Code are considered to be too lenient.
  (c) Magnetic particle
    (1) This method is preferred for linear flaws on or within 1/8 in of the surface, in materials that can be magnetized. To be effective, the magnetic field must be oriented so it crosses the flaw at an angle of roughly 45°.

Fortunately, most flaws in new impellers are longitudinally oriented with respect to the weld. Fatigue cracks in an impeller that has been in service might have random orientation, so that the magnetic field should be applied in two directions, roughly 90° apart. It is possible to detect an open gap under a vane where the fillets do not have full penetration. This is especially so if the throat of the weld is undersize or the gap is excessively wide. If these indications are strong (heavy), the inspector must satisfy himself that the weld is acceptable. In such a case, it might be necessary to weld a mock-up with a known flaw of the type suspected. A crack-like flaw will give a sharper-edged indication. The magnetizing force should meet or exceed ASME Code, Section VIII requirements.

(2) All cracks and crack-like flaws are to be addressed as stated earlier in this segment of our text. Any porosity indications should be judged the same as those disclosed by liquid penetrant examination.

(d) Ultrasonic examination: If the vendor opts to use this inspection method, the owner-purchaser should require the following:

(1) The shear wave of the weld to determine the degree of weld penetration and detect flaws per ASME Code, Section VIII. A straight beam can be used on fillet welds. UT examination is not used routinely on welded impellers. It has been used for special applications such as checking for under-bead cracks and on-plug welds, for lack of root penetration.

Examination of the fillet welds joining the vane to either the disc or cover can present a practical problem. These problems become more acute when the fillets, as is usually the case, do not have complete penetration. The difficulties are:

- Flaw orientation. Tight subsurface throat cracks and lack of penetration at the root of the weld may not be found by ultrasonic testing (UT).

- Small clearances in the gas passages. Usually they do not permit use of the crystals inside the passages. This requires any examination be done through the disc or cover.

- Varying material thicknesses. The discs and covers usually taper in thickness from the hub toward the periphery. Crystal movement must be adjusted to compensate for this.

- UT response from the open root of the tee joint makes interpretation difficult and confusing. With this type of joint only the area underneath the toe of fillets in the discs and covers can be confidently tested. More response is obtained if the weld has complete penetration through the vane.

(2) Ordinarily, the impeller manufacturer's standards for flaw acceptance and instrument calibration can be used. As a guide, whenever difficulty is being experienced with under-bead cracking, all flaws with an indicated depth and more than 1/8 in long are cause for rejection.

3. Repairs

(a) If the examinations show defective welds, etc, the impeller must be repaired, reexamined as before, and centrifuged again. This step is then followed by any final NDT required.

(b) If the material air hardens in response to the heat input from welding and will require preheat, it requires maintaining a defined interpass temperature and PWHT (Post-Weld Heat Treatment). These requirements must be met, regardless of impeller age. The inspector should not accept a repair on an impeller that was not made in accordance with the welding procedure used when the impeller was originally built, unless it is specifically approved by the owner's machinery engineer.

(c) Dimensional checking: Dimensional checking is required for impeller hub bore, outside diameter, eye diameter, vane width, and disc and cover thickness. Experience shows that special care is needed not to allow unacceptable tapering, or out-of-roundness of impeller bores on impellers that have been removed from a preexisting rotor. Dimensions must be within drawing tolerances or in accordance with documented engineering instructions superseding these drawings. They should be recorded for comparison with measurement made after overspeed testing. The owner-purchaser's designated inspector should witness these checks.

(d) Overspeed testing: Impellers of proven design should be overspeed-tested at 115% of maximum continuous, new impeller designs should be overspeed-tested at 120% of maximum continuous. All tests should be witnessed. In uprate situations the overspeed test speeds may differ from those originally used in the impellers and rotors initially sold to the owner-purchaser. With few exceptions a "scope of supply" or similar specification document should be developed and referenced.

After the overspeed test, each impeller should be visually examined again and any required NDT examinations should be witnessed. The owner's inspector should note that the points of highest stress are in the cover close to the eye of the impeller near where the vanes terminate. These are points where indications of possible failure will first show.

- API 617 assumes the compressor manufacturer has established its own acceptance standards for flaw indications. For casting and forging flaws, these standards can be compared with those given in the ASME Code, Section VIII, Division I.

- If the manufacturer's standards are more lenient than those listed in the ASME documents, the inspector should request instructions from the owner's machinery engineer, unless the owner-purchaser's scope or specification document has already covered the deviation.

- If repairs are necessary, the repaired area must be reexamined by the specified NDT method and the wheel overspeed tested again.

- Impeller diameters, including hub bore should be rechecked. If the growth exceeds the impeller manufacturer's tolerances, the wheel should be rejected by the inspector and the manufacturer's proposed action referred to the owner's machinery engineer for approval.

4. Rotor Inspection

   (a) The major components of the rotor assembly are the shaft, spacers, impellers, balancing drum, and thrust collar.

   (b) If, in uprate situations, the shafts that have originally run in the owner's compressor are being reused, ultrasonic testing of the shaft will not be required.

   (c) The critical shaft dimensions are the diameters over which shrink fits will be made, where keys will be placed, and at the journals. These dimensions must be carefully checked and recorded. The finish of the journals and probe surfaces can be examined again when run-out of the rotor is checked.

   (d) Any potential proposal to correct an undersized journal or shaft area by chrome plating cannot be approved by the inspector; this must be done by the owner's machinery engineer.

(e) The surface preparation and other parameters may have to be closely investigated by knowledgeable third parties. The authors' basic position has been to not accept plating as a repair for increasing shaft diameters. However, there have been cases where approval was given to add 10 to 15 mil (0.010–0.015 in) to the shaft diameter.

(f) When the impellers are assembled on the shaft with a shrink fit, the inspector should verify that the manufacturer's personnel control the bore diameters of the hubs and the temperature to which the impellers are being heated.

(g) Before witnessing the final balance, the owner's inspector should review shop assembly records and review the interference fits of wheels on shafts against the manufacturer's standards. Normal interference is 0.001 in/in of shaft diameter.

(h) High-speed dynamic balancing of the compressor rotors is required and the final balance check must be witnessed. Balancing procedures and balance tolerances are described in API 617. The version (or late edition/revision) that applied at the time of original purchase is usually accepted by the owner-purchaser.

(i) It should be noted, however, that the maximum balancing and proof speeds may have changed in uprate situations. The owner's inspector should always verify that, when an incremental balancing procedure is required by the compressor manufacturer's engineering personnel, the required procedure is actually followed by the manufacturer's shop personnel.

5. Run-Out Checks of the Assembled Rotor

(a) These run-out checks should be witnessed. The run-out check made after rotor assembly is particularly important because the measurements will indicate if the rotor has been assembled properly or has bowed due to stresses introduced during assembly or by mishandling.

(b) For a run-out check, the rotor can be supported on level knife edges or checking can be done while still in the balancing machine. A dial indicator is set up on the diameter to be checked and the rotor is rotated. The total reading is the run-out.

(c) Run-out checks should be made on the bearing journal surface, the radial vibration probe surface, impeller eyes, and thrust collar surfaces.

(d) These readings are to be compared to those on shop assembly drawings. Any measurements outside of tolerance must be questioned, as there may be bowing of the shaft or assembly errors. The owner's inspector must be certain that the mechanical run-out at the radial vibration probe surfaces does not exceed 0.2 mil (0.0002 in). The inspector must also verify that the shaft surface finish at radial probe locations is equal to the finish on the journals. Axial probe-sensing surfaces must be perpendicular to the shaft axis within 0.2 mil (0.0002 in).

(e) Next, the electrical run-out in the eddy current probe areas of the shaft must be checked and recorded. If the total (mechanical plus electrical) run-out exceeds 0.25 mil on new shafts, or 0.5 mil (0.0005 in) on reused shafts. The surface must be burnished or, if it will not clean up, be fitted with a sleeve. Final compliance must be verified by the owner's inspector.

(f) At this stage in the manufacturing cycle, it must be verified that the residual magnetism in shafts and impellers does not exceed 3 Gauss (3 G). Do not overlook this requirement!

6. Safety Issues: The inspection work described above involves close visual examinations, witnessing of tests, and, at times, the use of gages to ascertain the accuracy or correctness of the manufacturer's quality control effort.

(a) The minimum eye protection required while engaged in this type of work is safety spectacles with side shields. Hearing protection may be required in certain areas of the manufacturer's plant. As a guide, if conversation is difficult due to noise level, use hearing protection. Beware of damage to fingers while inspecting impellers. Do not examine equipment while suspended from a crane.

(b) The inspector must not wear ties, dangling decorations, loose fitting clothing, or loose long hair while working around rotating machinery.

What We Have Learned

You get what you inspect, not what you expect. Other than bearings and seals, impellers are the most likely process gas compressor parts to fail. The majority of these compressor impellers ("wheels") are custom designed and custom fabricated. Welded construction is widely used.

Impeller fabrication and testing require quality control inspection. While competent compressor manufacturers have quality control departments, it would be unrealistic to expect that their judgments always favor the user-purchaser. Moreover, "downsizing" and "right-sizing" tends to disproportionately affect quality control and inspection departments.

The owner's inspector need not be an expert in all the techniques used in the various manufacturing and testing practices, but needs to be an expert in asking the right questions. A competent vendor or manufacturer will answer these questions without reluctance. That is because such a manufacturer will have nothing to hide and, by answering the owner-purchaser's questions, will fulfill a much-needed tutoring and mentoring role.

Reference

1. Barney McLaughlin, Hickham Industries, Correspondence with the Authors, LaPorte, TX, 1983.

CHAPTER 8

Compressor Maintenance and Surveillance Highlights

A later chapter (Chapter 10) will make the point that maintenance cost avoidance can be built into a compressor specification. Maintenance cost avoidance is ascertained during "before-purchase machinery quality assessment," or *MQA*. At all times, specifying, purchasing, installing, and operating reliable compressors makes more sense than buying compressors on faith, past reputation, or lowest bid price.

When purchasing new compressors the best available solution is specify, design, built, and install machines with low risk in mind. The need for frequent or specific types of maintenance can be "designed out;" the most desirable compressor is designed for uptime extension, low failure risk, and lowest possible life cycle cost. On equipment that is already in service one has to opt for the next-best available solution. This would mandate that every maintenance intervention is viewed as an opportunity to upgrade. If upgrading is feasible, it will also have to be cost-effective. Upgrading should then result in the systematic reliability improvement of previously weak links in the component chain.

A prominent, but dated source[1] cites 13% of all failures of turbocompressors as being due to errors or omissions in condition monitoring and maintenance. With the advance in monitoring technology and modern operating and maintenance practices one would assume that this general number might not be as high today. What then are good monitoring and maintenance practices around turbocompressors?

Compressor condition monitoring has the following components:

1. Proper response to supervisory instrumentation such as alarms and trips.

2. Periodic observation and evaluation of operating parameters such as the compressor physical condition and its performance efficiency. This would include measuring and judging the rate of deterioration of mechanical and performance conditions for input into maintenance plans. Vibration analysis and aerodynamic performance calculations come to mind. Daily compressor operator rounds should be structured following the principles of operator-driven reliability, "ODR".[2]

3. Evaluation of operating trends. This should include auxiliary systems, such as lubrication and seal oil consoles, compressor on-line washing facilities, and dry gas seal support systems.

4. Periodic testing of lubrication and seal oils. Six basic analyses are required: Appearance test, testing for dissolved water, flash point test, viscosity test, the determination of the total acid number (TAN), and the determination of the additive content.[3]

5. Periodic testing of emergency shutdown devices (ESD) and other fail-to-danger components, such as exercising the compressor's surge control valve loop and the trip and throttle (T&T) valve on steam turbine-driven compressor trains.

6. Data logging and automated record keeping such as the number of unplanned trips per train per year as a basic indication of compressor reliability.

7. Diagnosis of problems, appraising their severity and deciding what action to take.

8. Remedial action and execution planning.

9. Corrective measures should preferably be applied on-stream to reduce the impact on compressor availability. On-line flushing (washing) would be a good example.

Generally, turbocompressors have maintenance inspections, overhauls and repairs (MIO&R), elsewhere called *IRD*, meaning inspection and repair downtime. These terms are used interchangeably with "turnarounds." MIO&Rs or IRDs are scheduled in periodic intervals ranging from 2 to 10 years, depending on the type of service. Maintenance intervals in clean services in the hydrocarbon industries of 6 to 10 years are not uncommon. The extent of MIO&R efforts ranges from simple bearing inspections to opening the compressor and replacing the rotor with a spare rotor drawn from specialized spare parts storage. Used rotors are examined for rubs at labyrinth seal locations and for fissures and cracks around impeller eyes on radial compressors. Rotor blades and also stator blades of axial compressors require thorough examination and testing. In all cases nondestructive test (NDT) procedures are being applied.

As the scheduled compressor turnaround approaches, it is best practice to review the machine's operating and maintenance history. Then, if there are any defects noted at inspection, it would be well to ask the following questions:

1. Are any of these defects repeat occurrences?
2. If so, can they be expected at this turnaround?
3. What steps can be taken to eliminate them?
4. What action should be taken at this time?

A thorough pre-turnaround review should be undertaken in order to plan the work required. It should consist of:

1. An assessment of the compressor's mechanical condition
2. A performance check
3. A diligent review of the machine's past history

Preventive and Predictive Maintenance Explained

Still, both the good and not-so-good compressors will have to be maintained, and the prevailing maintenance strategies can be either preventive or predictive. *Preventive maintenance (PM)* is time based whereas *predictive maintenance (PdM)* has as its goal compressor operation until detectable defects start to develop. Detection requires high-quality predictive monitoring methods and comes at a cost. Qualified personnel have to be employed and ongoing monitoring efforts have to be quite precise. Management expects PdM to determine when a failure will occur and then plan an outage accordingly.

It is fair to say that certain state-of-art predictive routines can be used to minimize the impact of a premature failure, or to understand when a machine drifts into off-design operation. But cost savings always come back to the knowledge ingredient. None of the various PdM judgments can be made without experience. Real depth of experience and wisdom will be needed when several seemingly minor deviations occur and converge.

Preventive maintenance encompasses periodic inspection and the implementation of remedial steps to avoid unanticipated breakdowns, production stoppages, or detrimental machine, component, and control functions. Predictive, and to some extent also preventive maintenance, is the rapid detection and treatment of equipment abnormalities before they cause defects or losses. This is evident from considering lube oil changes. This routine could be labeled preventive if time-based, and predictive if done only when testing shows an abnormality in the properties of the lubricant. Without strong emphasis and an implemented preventive maintenance program, plant effectiveness and reliable operations are greatly diminished.

In many process plants or organizations, the maintenance function does not receive proper attention. Perhaps because it was performed as a mindless routine or has, on occasion, disturbed well-running equipment, the perception is that maintenance does not add value to a product. This may lead management to conclude that the best maintenance is the least-cost maintenance. Armed with this false perception, traditional process and industrial plants have underemphasized preventive, corrective, and routine maintenance. Many have neglected to properly develop maintenance departments, elected not to pursue proper training of maintenance personnel, and to not optimize predictive maintenance. Many unforeseen compressor failures and safety hazards have resulted from not understanding what this is really all about.

Correctly executed, maintenance is not an insurance policy or a security blanket. It is a requirement for success. Without effective preventive maintenance, equipment is certain to fail during operation. To be effective, maintenance must be selective and has always had to be selective. Selective preventive maintenance (selective PM) results in damage avoidance, whereas effective PdM allows existing or developing damage to be detected in time to plan an orderly shutdown.

Compressor Maintenance in Best Practices Plants

Four levels of effective compressor maintenance exist. Although there is some overlap, the levels of maintenance are:[1]

1. *Reactive, or breakdown maintenance.* This type of maintenance includes the repair of equipment after it has failed, in other words, "run-to-failure." It is unplanned, unsafe, undesirable, expensive, and, if the other types of maintenance are performed, usually avoidable.

2. *Selective preventive maintenance.* Selective preventive maintenance includes lubrication and proactive repair. On-stream lubrication of, say, the admission valve control linkage on certain steam turbines should be done on a regular schedule. In this instance, anything else is unacceptably risky and inappropriate.

3. *Corrective maintenance.* This includes adjusting or calibrating the equipment. Corrective maintenance improves either the quality or the performance of the equipment. The need for corrective maintenance results from preventive or predictive maintenance observations.

4. *Predictive maintenance and proactive repair.* Predictive maintenance predicts potential problems by sensing operation of equipment. This type of maintenance monitors operations, diagnoses undesirable trends, and pinpoints potential problems. In its simplest form, an operator hearing a change in sound made by the equipment predicts a potential problem. This then leads to either corrective or routine maintenance. Proactive repair is an equipment repair based on a higher level of maintenance. This higher level determines that if the repair does not take place, a breakdown will occur.

Predictive maintenance instrumentation is available for both positive displacement and dynamic compressors. It exists in many forms and can be used continuously or intermittently. It is available for every conceivable type of machine and instrumentation schemes range from basic, manual, and elementary, to totally automatic and extremely sophisticated. Recommended instrumentation depends on compressor size and owner's sparing philosophies. As an example, a facility may opt to install three 50 percent machines, two 100 percent machines or perhaps only one 100 percent machine in a given service. Moreover, unless the value of downtime avoidance is quantified it will not be possible to make firm recommendations as to the most advantageous level of monitoring instrumentation, shutdown strategies, etc.

There are many competent manufacturers of manual monitoring equipment and manual monitoring is often used on small air compressors. Advanced predictive maintenance on-stream systems are generally used with large process compressors to continuously monitor vibration behavior. By gathering vibration data and comparing these data with normal operating conditions, both manual and continuous systems can predict and pinpoint the cause of a potential problem. The trouble is that detecting vibration is different from eliminating vibration.

An intelligent but highly *selective* preventive maintenance program may lead to actions that prevent bearing distress and thus prevent vibration from occurring in the first place. A *selective preventive maintenance* program may well be a more cost-effective program than any program or strategy that waits for defects to manifest themselves. This fact establishes that sweeping management edicts that disallow *all manner of preventive maintenance* on compressors do not harmonize with the principles of asset preservation and best practices.

Traditionally, industry has focused on breakdown maintenance, and unfortunately, many plants still do. However, in order to minimize breakdown, maintenance programs should focus on levels 2 through 4.

Emergency Repairs Should Be Minimized

Plant systems must be maintained at their maximum level of performance. To assist in achieving this goal, maintenance should include regular inspection, cleaning, adjustment, and repair of equipment and systems. Repair events must be viewed as opportunities to upgrade. In other words, the organization *must* know if upgrading of failed components and subsystems is feasible and cost-justified. On the other hand, performing unnecessary maintenance and repair should be avoided. Breakdowns occur because of improper equipment operation or failure to perform basic preventive functions. Overhauling equipment periodically when it is not required is a costly luxury; upgrading where the economics are favorable is absolutely necessary to stay in the forefront of profitability.

Regardless of whether or not PdM routines have determined a deficiency, repairs performed on an emergency basis are three times more costly in labor and parts than repairs conducted on a pre-planned schedule. More difficult to calculate, but high nevertheless, are costs that include shutting down production or time and labor lost in such an event.

Bad as these consequences of poorly planned maintenance are, much worse is the negative impact from frequent breakdowns on overall performance, including the subtle effect on worker morale, product quality, and unit costs.

Effectiveness of Selective Preventive Maintenance

Selective preventive maintenance, when used correctly, has shown to produce considerable maintenance savings. Sweeping, broad-brush maintenance, including the routine dismantling and reassembling of compressors is wasteful. It has been estimated that one out of every three dollars spent on broad-brush, time-based preventive maintenance is wasted. A major overhaul facility reported that "60 percent of the hydraulic pumps sent in for rebuild had nothing wrong with them." This is a prime example of the disadvantage of performing maintenance to a schedule as opposed to the individual machine's condition and needs.

However, when a *selective* preventive maintenance program is developed and managed correctly, it is the most effective type of maintenance plan available. The proof of success can be monitored and demonstrated in several ways:

- Improved plant availability
- Higher equipment reliability
- Better system performance or reduced operating and maintenance costs
- Improved safety

A plant staff's immediate maintenance concern is to respond to equipment and system functional failures as quickly and safely as possible. *Every maintenance event must be viewed as an opportunity to upgrade so as to avoid repeat failure.* This italicized sentence is the key to superior maintenance. Even good maintenance refers to relatively frequently scheduled work. Systematic upgrading will extend allowable intervals between shutdowns.

Know Your Existing Program

The starting point for a successful long-term selective maintenance program is to obtain feedback regarding effectiveness of the existing maintenance program from personnel directly involved in maintenance-related tasks. Such information can provide answers to several key questions, and the answers will differ from machine to machine and plant to plant. Your in-plant data and existing repair records will provide most of the answers to the seven questions given below. A competent and field-wise consulting engineer will provide the rest:

1. What is effective and what is not?
2. Which time-directed (periodic) tasks and conditional overhauls are conducted too frequently to be economical?
3. Which selective preventive maintenance tasks are justified?
4. What monitoring and diagnostic (predictive maintenance) techniques are successfully used in the plant?
5. What is the root cause of equipment failure?
6. Which equipment can run to failure without significantly affecting plant safety and reliability?
7. Does any component require so much care and attention that it merits modification or redesign to improve its intrinsic reliability?

It is just as important that changes not be considered in areas where existing procedures are working well, unless some compelling new information indicates a need for a change. In other words, it's best to focus on known problem areas.

To ascertain focus, continuity of information, and proper activities relative to maintenance of plant systems, some facilities assign responsibility for well-delineated plant systems to a knowledgeable staff person. All maintenance-related information, including design and operational activities relating to such a system are funneled through this expert. He or she refines the maintenance procedures for the systems under his jurisdiction and reshapes preventive maintenance into selective maintenance.

Maintenance Improvement

Problems associated with machine uptime and quality output affect several functional areas. Many people, from plant manager to engineers and operators, make decisions and take actions that directly or indirectly affect machine performance. Production, engineering, purchasing, and maintenance personnel as well as outside vendors and stores use their own internal systems, processes, policies, procedures, and practices to manage their sections of the business enterprise. These organizational systems interact with one another, depend on one another, and constrain one another in a variety of ways. Some constraints are appropriate; other constraints can have disastrous consequences on equipment reliability.

That said, program objectives need to be clearly defined. An effective maintenance program should meet the following objectives:

- Unplanned maintenance downtime does not occur.
- Equipment condition is known at all times.
- Where justified, preventive maintenance is performed regularly and efficiently.
- Selective preventive maintenance needs are anticipated, delineated, and planned.
- The maintenance department performs specialized maintenance tasks of the highest quality.
- All craftsmen are highly skilled and participate actively in decision-making process.
- Proper tooling and information are readily available and being used.
- Replacement parts requirements are fully anticipated and components are in stock.
- Maintenance and production personnel work as partners to maintain equipment.

Following these general guidelines for centrifugal and reciprocating compressors will give positive results.

What We Have Learned

The main lesson we learned decades ago is that one deviation alone might not be enough to bring on a compressor failure, but when several more deviations combine, the failure risk increases exponentially. So, while it might be possible to avoid more serious failures by implementing an automatic compressor-unloading scheme, or adding bells and whistles that annunciate excessive temperatures and vibrations and seal deficiencies and the like, there will never be any substitute for the human brain supplying both logical root cause failure analysis and up-front failure prevention processes.

Achieving these up-front processes before calamities and finger-pointing occur requires both training and accepting accountability. The operator, supervisor, or manager accepting a deviation from established practice should somehow be motivated (or should be compelled) to understand the potential ramifications of bypassing or not following established guidance and should document this understanding in writing. As the documentation requirements are then enforced, fewer deviations would be tolerated. Accept the initial incremental cost outlay needed to do things right. The apparent expenditure of time and money will ultimately bring rich rewards in safety, reliability, and increased profitability.

If you're stuck with an existing compressor, view every maintenance event as an opportunity to ask if upgrading is feasible. Suppose the answer is yes and it can be shown to be cost-justified, *do it*. It's your professional duty to the stakeholders. Stakeholders are not just investors, they're husbands and fathers and families.

References

1. Bloch, Heinz P., and J. J. Hoefner, *Reciprocating Compressors: Operation and Maintenance*, Gulf Publishing Company, Houston, TX, 1998.
2. Bloch, Heinz P., "Operator-driven reliability," *Maintenance Technology*, April 2008.
3. Bloch, Heinz P., and F. K. Geitner, "Machinery failure analysis and troubleshooting," 4th ed., Elsevier Publishing, New York/London/Tokyo, 2012.

CHAPTER 9

Inspection and Repair Guidelines for Rotors

To be reliable, reproducible, and to give continuity to compressor repair, compressor owners must adhere to procedures. Continuity means passing on knowledge to successive generation of workers so as not to repeat errors.

In general, the manufacturer can provide the details that cover the routine maintenance inspection (during plant turnarounds), or the disassembly, stacking, and balancing of centrifugal compressor rotors if performance enhancing or throughput increasing ("uprating") are involved.

A preexisting procedure serves as the usual framework of detailed repair instructions. Detailed instructions must be developed from standardized or preexisting procedures. These detailed instructions are then issued on an individual and as-required basis.

For the purpose of this text and to be concise, we are melding inspection and repair guidelines into one multifaceted document. From this starting-point document the owner-purchaser may then proceed in a variety of ways; the most prevalent are:

- Separate these multifaceted guidelines into stand-alone documents and use the stand-alones for follow-up. Stand-alones could be used in work performed onsite or offsite, or by the owner-purchaser's permanent employees, or by another entity.

- Use these guidelines as a discussion basis with either the OEM or a non-OEM repair facility. Use these guidelines to reach agreement on deliverables.

Compressor owner-purchasers often deal with issues that must be assessed in conjunction with original procurement. At other times, owner-purchasers are involved in compressor uprates whereby a rotor is being reworked by the OEM or by a competent non-OEM to meet new performance parameters. Many of the data found below are relevant to both original procurement and future work situations; this text combines diverse data. Again, the owner-purchaser may opt to separate the data into one of many different scope and procedures documents.

Phase I (Initial Preparation)

Preliminary Work Lists and Tabulations of Deliverables for Work Performed during Planned Shutdowns

1. The compressor owner must have carefully prepared a comprehensive list of parts to be on hand and parts to be replaced during compressor shutdown.

2. The parts on hand must be measured and correct materials and dimensions verified against vendor drawings.

3. This verification is needed because an estimated 5 percent of spare components are flawed for one reason or another and will have to be reworked or replaced to be suitable for use in the compressor. In one documented case, 14% of all spare parts were found in need of rectification before a major turnaround was commenced.[1]

4. While the equipment is still in operational service and approximately 9 months before a planned shutdown, a list of *likely defective* parts in the equipment must be generated. This list must be compiled on the basis of predictive maintenance (PdM) monitoring and preliminary root cause failure identification meetings. A vendor's field engineer may have to be hired to do the verification work described in (1); this field person must bring along whatever proprietary vendor's drawings are not otherwise available at the user's site.

Inspection

1. Careful visual inspection must be performed after removing the top half of the compressor or, in the case of barrel-type construction, after the rotor is removed from the machine. Typical deviations are shown on a sketch (Fig. 9.1).

   • Photograph the rotor in its "as received" condition. In addition, close-up color photos are to be taken of any unusual or abnormal condition, and a photo log should be maintained for all work performed. Identification of all

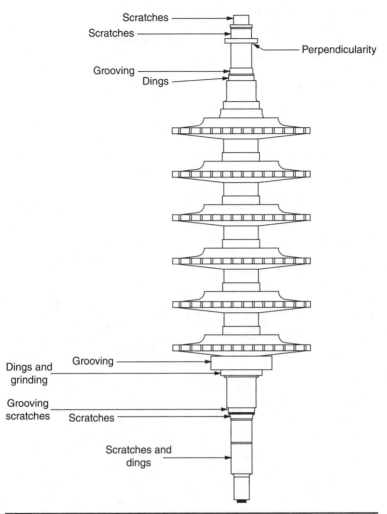

FIGURE 9.1 As part of "incoming inspection" the compressor manufacturer or rebuild shop identifies all discrepancies and documents the "as received" condition of a centrifugal compressor rotor.

items, including equipment number and part name, should be clearly shown on all photographs.

- Take samples of residues and deposits.
- Photograph any unusual appearances and note abnormal looseness or tightness.
- Also, measure and document dimensions from a distinct shaft feature (such as a shaft shoulder etc, Fig. 9.2) to various important locations along the rotor axis.

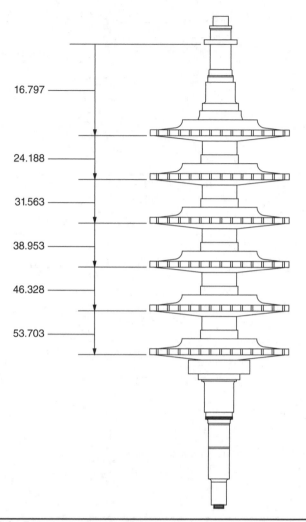

16.797

24.188

31.563

38.953

46.328

53.703

Figure 9.2 Typical mapping of axial distances on a six-stage rotor.

- Measure and record all diameters as shown in Fig. 9.3. (Note: All micrometers must be checked for correct calibration against a precision standard prior to use.)
- As clearly defined in a job scope or instruction document, a designated member of the owner's workforce, the OEM vendor-manufacturer, or the designated outside-repair contractor should make two electronic sets of standard color photos.
- The distance to the centerline of the vibration probe area should be recorded on each end of the rotor (Fig. 9.4) to

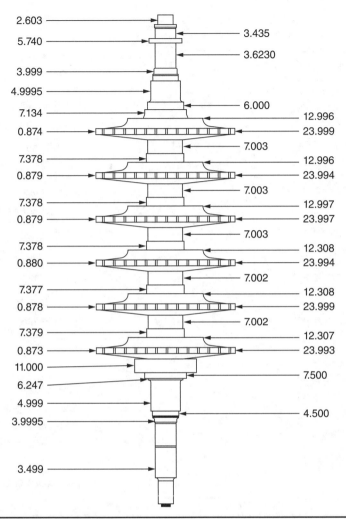

2.603
5.740
3.435
3.6230
3.999
4.9995
7.134
6.000
0.874
12.996
23.999
7.003
7.378
12.996
0.879
23.994
7.003
7.378
12.997
0.879
23.997
7.003
7.378
12.308
0.880
23.994
7.002
7.377
12.308
0.878
23.999
7.002
7.379
12.307
0.873
23.993
11.000
7.500
6.247
4.999
4.500
3.9995

3.499

Figure 9.3   Typical mapping of important diameters on a six-stage rotor.

allow quality control of runouts during repair. Afterward, the probe areas must be adequately protected from damage such as rusting or scratches that could otherwise occur before and during the rotor repair.

Cleaning

Prepare to clean the rotor to remove dirt, rust, and other foreign material. (Important: Do not use steam to clean the assembled rotor.)

- Protect all bearings, seal, probe, and coupling surfaces before and during cleaning.

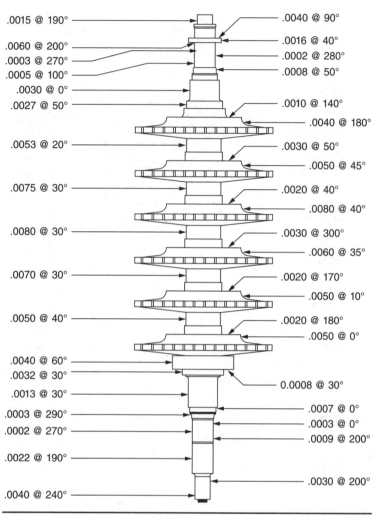

.0015 @ 190°
.0060 @ 200°
.0003 @ 270°
.0005 @ 100°
.0030 @ 0°
.0027 @ 50°

.0040 @ 90°
.0016 @ 40°
.0002 @ 280°
.0008 @ 50°

.0010 @ 140°
.0040 @ 180°

.0053 @ 20°

.0030 @ 50°
.0050 @ 45°

.0075 @ 30°

.0020 @ 40°
.0080 @ 40°

.0080 @ 30°

.0030 @ 300°
.0060 @ 35°

.0070 @ 30°

.0020 @ 170°
.0050 @ 10°

.0050 @ 40°

.0020 @ 180°
.0050 @ 0°

.0040 @ 60°
.0032 @ 30°
.0013 @ 30°
.0003 @ 290°
.0002 @ 270°
.0022 @ 190°
.0040 @ 240°

0.0008 @ 30°

.0007 @ 0°
.0003 @ 0°
.0009 @ 200°

.0030 @ 200°

FIGURE 9.4 Typical mapping of runout readings on a six-stage rotor.

- Blast-clean with (1) glass beads, (2) walnut shells, or (3) grit, 200 mesh or smaller.

- Coat all surfaces with a film of oil or other protective coating after cleaning.

- Unless otherwise specified in the job scope, use applicable nondestructive examination procedures (Magnaglo magnetic particle testing, or Zyglo liquid penetrant inspection) to determine the existence and location of any defects such as cracks on the rotor. Record the size, location, and orientation of any defects on a sketch. (Note: This step will not normally be done on rotors that must be disassembled.)

Deficiency Mapping

Additional deficiencies may become evident after cleaning. The owner's engineer must now fine-tune and update the sketch in Fig. 9.1.

1. Note on the same Fig. 9.1 or on a similar sketch the size, location, and orientation of any rubs, erosion, corrosion, or other damage resulting in loss or displacement of material and/or buildup of deposits, including any in keyways or under impellers or sleeves. If build-up is substantial, remove a sample and take steps to have it analyzed, as needed. (Do not overlook the possibility of discovering more damage as you proceed. If deposits are suspected to have caused damage such as cracks, corrosion, etc, the repair work may have to be delayed until complete analysis is performed.)

2. Take additional photographs and relate them to the various sketches made earlier.

3. Regardless of who will have involvement in, or responsibility for the work, photograph the rotor in its "as received" condition. In addition, close-up color photos are to be taken of any unusual or abnormal condition, and a photo log should be maintained for all work. Identification of all items, including equipment number and part name must be clearly shown on all photographs. Unless otherwise specified in the jobs scope, make two (2) electronic copies of all photographs taken.

4. Measure and record all relevant and accessible dimensions of the rotor as received on a checklist worksheet designed for the particular rotor. (Note: All micrometers must be checked for correct calibration against a precision standard prior to use.)

   (a) Impeller outside ("G") and suction eye diameters ("D") (Fig. 9.5).

   (b) Seal sleeves, spacers, and other running clearance regions (Figs. 9.6 and 9.7).
   Journal and seal area diameters (check both ends and center of each area for roundness and taper, Figs. 9.4 and 9.8). Record all dimensions.

   (c) Coupling fits, Fig. 9.8 (if tapered, record minor and major diameters, length of taper, and percent of contact area, which is blued using a ring gage). Blade-type micrometers are preferred for tapered coupling fits. The stand-off dimension of the ring gage must also be recorded.

   • Depth, length, location, and type of any coatings, overlays, etc.

   • Gaps between all adjacent shrunk-on parts.

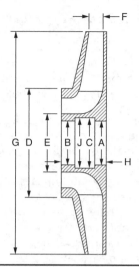

Figure 9.5 Typical impeller with important dimensions.

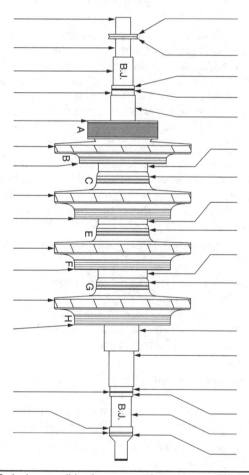

Figure 9.6 Typical accessible clearance regions that must be measured.

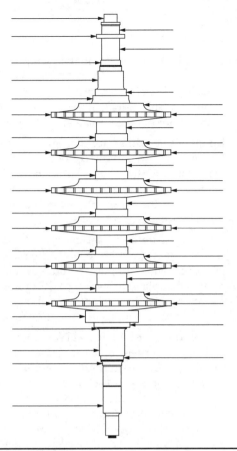

Figure 9.7 All accessible clearance and fit-up regions must be measured.

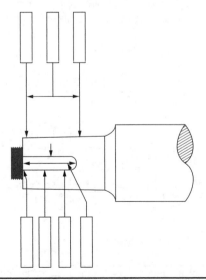

Figure 9.8 Important coupling fit and keyway dimensions.

- Check all runouts with the shaft supported at the bearing journals on "Vee" blocks. (Note: Vee block lengths must be at least equal to one-half of the bearing journal diameters. The entire length has to be used for support in the center of the journal, labeled B.J. in Fig. 9.6.)

- Runouts (including vibration-probe areas) should be logged-in relative to the coupling's (driven end) keyway centerline; this centerline is labeled the zero-phase reference. If the coupling area is double keyed or has no keyway, the thrust collar keyway will be used as the zero reference.

- If this is also not possible, an arrow should be stamped on the end of the shaft to show the plane of the zero-phase reference. Runouts should be recorded as viewed from the coupling (driven) end of the rotor, while rolling the rotor in the direction of normal rotation (i.e., phase angle increasing opposite normal rotation).

- Check and record mechanical runout on shaft fit areas, seal areas, journals, probe target areas, and any other running clearance areas. As a minimum, axial runout must be recorded for each impeller suction eye and on both sides of the thrust collar (or on the thrust collar shaft shoulder if the thrust collar is removed). Mechanical runout check locations and diameter measurements on compressor shaft could be combined, Fig. 9.9.

- Check and record vibration probe area runouts using a calibrated eddy current probe transducer (minimum instrumentation requirement for this type of runout

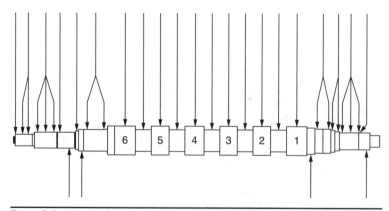

Figure 9.9 Mechanical runout check locations and diameter measurements on compressor shaft could be combined.

measurements). Prior to checking runouts in the probe region, record a graph of gap in 10 mil (0.010 in) versus transducer output voltage scale on the target area. There could possibly be more than one such target areas due to different metallurgical composition.

- On target areas less than 4 in in diameter, calibration must be performed radically on the shaft and appropriately noted on the calibration graph. Probe area runout tolerance is 0.25 mil (0.00025 in) of combined electrical and mechanical runout.

- Check and record axial stack-up dimensions, referenced from the thrust collar shaft shoulder (or active side of the integral thrust collar), to each impeller (the gas path surfaces nearest the thrust collar) and to all other major components that may be removed from the shaft. (There are no exceptions to this important requirement.)

- Weigh and record the total weight of the rotor.

- Check balancing must be done on a spare rotor being readied for operational use and on rotors that have simply been cleaned.

- If both bearing journals are round within 0.0003 in, perform a check balance of the rotor, using fully crowned half keys in all exposed keyways.

- Record amounts, radii, and locations of required temporary corrections, and weights of all half keys. (Note that check balancing is not required on rotors where damage had caused an obvious and large unbalance.)

  (When specified in the job scope or a related instruction document, dismantle the rotor completely. In that case, refer to Chapter 6 for details.)

5. Rotor Dismantling

   (a) During disassembly, check and record axial dimensions between the thrust collar shaft shoulder (or active side of the integral thrust collar) to the faces of the bores nearest the thrust collar on each impeller.

   (b) Visually inspect each part removed, and measure and record all shaft and component sizes on the appropriate checklist for the rotor. (Measure for roundness and taper on both ends of each component.) These measurements must include:

      - Bare shaft fit sizes, overall length, and runouts (including thrust collar shoulder).

- Impeller bore sizes, and lengths and widths of land fit areas.
- Bore sizes and lengths of shaft spacers, seal sleeves, etc.
- Balance piston bore size and length.
- Thrust collar and spacer bore sizes and lengths.
- Key clearances on all keyways. With keys installed in shaft keyways, measure from tops of keys to opposite side of shaft.
- Measure across bore of components to bottoms of keyways, then verify the dimensional compatibility of keys and keyway locations.

6. Nondestructive testing (NDT) or previously in service rotors

In general and unless otherwise specified in the job scope, the vendor or rebuild shop must use applicable nondestructive procedures (Magnaglo or Zyglo) after disassembly to determine the existence and location of any defects in the shaft and all component parts. The shaft should be ultrasonically tested in addition to the above. Record the size, location, and orientation of any defects on a sketch. Again, refer to Chapter 8 for more detail.

Be sure to take the following steps for including appropriate records and data:

(a) Photograph removed parts with abnormal or unusual conditions as described earlier in this chapter.

(b) Demagnetize all components and check and record residual magnetism on all parts of the rotor. Maximum allowable residual magnetism is 2.0 G, as measured with a digital Gauss meter and Hall-type probe.

(c) Request that owner's engineer be notified upon completion of Phase I. A copy of all Phase I documentation must be submitted for his use and records. Evaluate the results of Phase I and review or prepare the job scope for necessary repairs.

Phase II (Repairs)

The basis for such repair recommendations will be the minimum necessary to produce a safe reliable rotor that will perform efficiently for normal plant runs. Select a repair facility (OEM or non-OEM) that agrees to provide repair recommendations based on these criteria for review by owner's engineer prior to proceeding.

Phase III (Assembly and Balancing)

Owners' engineers should be involved in every decision that affects the long-term reliability of centrifugal compressors. They must determine if the manufacturer or repair facility adheres to standards that have long been maintained by reliability-focused users. While not every one of the various requirements may deserve to be rigidly enforced, the impact of deviating from them must be clearly understood. Therefore, except in situations where the manufacturer or repair shop has even more stringent requirements, the deviation should be the subject of discussion and mutual agreement. All conclusions must be documented in writing. Even the support mandrels used for impeller balancing must be precisely machined and balanced. Good mandrels contribute balance quality vibration-free operation to centrifugal compressor rotors.

1. Dynamically balance the bare shaft to tolerance $4W/N$ per plane, using fully-crowned half keys in all keyways. In this expression, W equals the weight of the workpiece in pounds, and N equals the maximum continuous speed of the compressor in RPM. The result is expressed as "ounce-inches" and further explanations can be found in balancing-related texts. In any event, when half keys are required, use keys that completely fill the keyway(s) and record the weight of each key.

    Half keys are not required if two conditions are met:

    (a) There are an equal number of keyways 180 degrees apart on each side of the shaft and no two adjacent keyways are in the same side of the shaft, and

    (b) All keyway lengths and depths are the same.

2. Set-up each individual impeller in a lathe or vertical boring mill and indicate the bore centerline true to within 0.0001 in.

3. Record all phase-related runouts on the suction eye face and periphery. Zero phase will be at the impeller keyway. If impellers are not keyed, zero phase must be stamped on the suction eye face.

4. All runouts must be recorded while turning the impeller in the direction of normal rotation and as viewed from the suction side. With the bore trued, the face of the suction eye should not run out in excess of 0.002 in total indicator reading (TIR).

5. Machine an indicator flat on the suction eye face if runouts exceed 0.002 in TIR.

6. Special instructions pertain to the mandrels used for impeller ("wheel") balancing. It is necessary to individually balance, on a balancing mandrel, each wheel and the balance piston. Tapered spring mandrels are not allowed.

7. Similarly, mandrels with measurable eccentricity noted by a dial indicator graduated in 0.0001 in increments are not acceptable to reliability-focused users. Before mounting any wheel, the mandrel itself should have a ground finish and be precision-balanced prior to cutting any keyways.

8. Balance tolerance in ounce/inches must be determined for each plane by the formula 4W/N. A keyway to accept the component's job key must then be cut in the finished mandrel.

9. Key top clearance should not be allowed to exceed 0.005 in. Each part balanced should have a 0.002 in minimum interference fit to the mandrel and a maximum interference fit equal to that between the impeller and the shaft.

10. Unless agreed by the owner's engineer, component and face runouts must duplicate those recorded in Step (b). Record phase-related runouts for each component mounted on the mandrel prior to balancing.

11. A residual unbalance test must be performed on the first of a series of like components for each different balance machine used. This typically consists of a minimum of six test points (60 degrees apart), with additional points possibly required to ascertain that minimum and maximum readings are obtained.

12. If balance machine readouts are found to be in error by more than 25 percent as verified by the residual test, a residual unbalance test should be performed and recorded for each component.

13. Normally, all impellers are being heated in a temperature-controlled oven for assembly to the shaft. Permission must be obtained from the owner's engineer prior to using torches for assembly.

14. On rotors that stack from the center out, allow stacking two (2) wheels or components at a time. On rotors that stack from one end, only stack one wheel or component at a time. All half keys shall be left in place in shaft keyways until it becomes necessary to replace them with the job keys.

15. At no time during the stacking process shall unbalance be induced by uneven distribution or lack of proper half keys. After each stacking step, allow all components to cool to 120°F or less.

16. Runouts should not change more than 0.0003 in TIR between component stacks, accounting for any phase differences, and must be recorded after each stacking step. The rotor must

again be turned in the direction of normal rotation during balancing, using the same plane of zero-phase reference that was used during runout checks.

17. Unless a particular compressor manufacturer or rebuild shop has reason and proven experience demanding otherwise, after each stacking step, the previously stacked impeller(s) should be "heat soaked" while rotating 300 to 500 RPM.

18. This is done by rapidly heating each impeller to 350°F to 400°F at the disc fit (closed side, also called "heel" in Fig. 9.5) only, using a "rosebud" torch. Be certain *not* to heat the shaft or the impeller cover fit (called "toe" in Fig. 9.5).

19. Allow to cool to 120°F or less.

20. Indicate impeller and shaft for runout. Maximum allowance is 0.001 in TIR on the shaft and 0.002 in TIR measured axially at the face of the impeller eye. Further, assembled component runouts must match those recorded in paragraph C within 0.001 in TIR accounting for any phase differences. If runout exceeds tolerance, then destack, check shaft runout, and finally restack.

21. Check the rotor balance after each stacking phase. If balance corrections are necessary, they must be confined to the last component(s) stacked only, without any corrections (temporary or otherwise) on other components. Continue this until the rotor stacking process is complete.

22. Unless otherwise specified and agreed to, the compressor manufacturer or repair shop should perform and record a residual unbalance after each wheel is mounted and balanced on the shaft. (In many agreements between owner and manufacturer, balancing of the first and last wheels is to be witnessed by the owner's engineer.)

23. Record keyway clearances during stack balancing as outlined in Phase I. Permissible top clearances should range from 0.003 to 0.006 in.

24. Axial clearance between each impeller spacer and balance piston should be 0.002 to 0.005 in.

25. Trim balance the final assembly to a balance tolerance of 4W/N (explained above as four times the journal static weight in each plane divided by the maximum operating speed).

Example 1   A wheel weighing 100 lb that operates at 10,000 RPM = $(100 \times 4)/10,000 = 0.04$ oz-in.

Example 2   A six-stage rotor weighing 1800 lb operating at 10,000 RPM with equal journal loading. It would equate to 1800 divided by two planes = 900 lb per plane. Thus, $(4 \times 900)/10{,}000 = 3600$ divided by 10,000 RPM = 0.36 oz-in (left plane) and 0.36 oz-in (right plane).

26. Perform and document a four (4)-point residual unbalance check for each end of the rotor after final trim balancing is complete.

27. Document all sizes, runouts (phase related to coupling keyway or other specified reference), fits, etc, and submit all original documentation with the final repair report within 3 days of job completion. For comparison purposes, final axial stack-up dimensions should be recorded as was done in Phase I, without exception. Unless otherwise specified in the job scope, final axial stack-up dimensions shall match those recorded during "as received" inspection within 0.015 in maximum deviation.

28. Check residual magnetism on the completed rotor. Maximum allowable is 2.0 G, as measured with a digital gauss-meter and Hall-type probe.

29. Check and record vibration probe area for electrical/mechanical runout as outlined earlier in Phase I. Make corrections, as necessary. Maximum allowable runout is 0.25 mil (0.00025 in) peak-to-peak, combined electrical and mechanical runout. Acceptable correction methods include burnishing, rolling, and micro-peening.

PHASE IV (Preparation for Storage or Shipment)

1. Coat the rotor thoroughly with a product suitable for a predefined storage or in-transit time. (Recall that steam is not allowed for cleaning the assembled rotor at any time.)

2. Wrap each probe target area separately, using MIL B-121 barrier material. Tape these areas and mark each with the words "Probe Area." Afterward, wrap the entire rotor using MIL B-121 barrier material.

3. Follow approved shipping or storage instructions.

4. If the rotor is to be shipped, the owner's engineer must approve the proposed shipping crate. Minimum material to be used: 4 in-by-4 in for the skid; 2 in-by-4 in for the box frame; ½ in-thick plywood for the sides and top.

5. The box should be marked *"Fragile, lift only while supporting the bottom."* The box should also be marked with the

owner-purchaser's company name and should be marked as the owner-purchaser's property.

6. The rotor must be supported rigidly at both bearing journals and in the center when crated. Acceptable materials for use between the wrapped journals and the supports are lead, PTFE, a high-performance polymer, or a Micarta-like product.

What We Have Learned

Existing rotors are first photographed when the compressor is opened at the owner's site. They are again very carefully mapped at the point of receiving, usually at a repair facility.

Each work procedure and repair step must be defined in writing. It is understandable that an owner-purchaser may not have the time or inclination to become a compressor designer; however, an owner-purchaser's staff member must ask questions that relate to a compressor repair. These questions must be answered by the OEM or non-OEM repair facility.

You should refuse to entrust your machine to a repair shop that is hesitant to describe its work processes or procedures. There should be no such thing as a proprietary repair or fabrication procedure that cannot be divulged to the compressor owner. It's your property; don't allow details to be hidden from the owner-purchaser's view or scrutiny.

Finally, remember again that you will get what you inspect. You will not get what you expect.

Reference

1. Bloch, Heinz P., *Improving Machinery Reliability*, Gulf Publishing Company, Houston, TX, 1982/1988/1998.

CHAPTER 10

Machinery Quality Assessment

Machinery quality assessment (MQA) is a multifaceted work effort aimed at uncovering risks and vulnerabilities that exist in low-cost or prototype-like compressor offers.[1] It has been estimated that MQA efforts require an up-front prestartup investment of 5% of the cost of equipment and will ultimately yield a 10-fold return on this incremental investment. Only experienced staff can make these assessments, and even they may need assistance from qualified third-party machinery specialists. Together, the purchaser's MQA staff and the outside contractors reach consensus and report their findings to the project manager.

Best-of-class companies use MQA on critically important compressors and drivers. The MQA effort consists of structured and definable reviews of drawings, calculations, and other documentation. MQA tasks and pursuits are separated into three phases.

Phase 1 efforts take place before a purchase order is issued; they are concentrating on vendor capabilities and specification development. The plant's operating philosophy, its location, available labor pool, cost of a day's outage, desired plant availability, and a host of other considerations determine which vendors are invited to bid and what should be the extent of supply. The value of potential preinvestment in future capacity expansion is among the various issues defined during Phase 1.

Phase 2 efforts are generally called design audits. They are structured endeavors that take place approximately 6 weeks after a purchase order has been issued. A design audit is conducted at the compressor vendor's design and manufacturing location and concentrates on compressor details that are only obtainable from near-finalized design drawings or calculations.

As an example, the audit may uncover that the vendor intends to use a thrust bearing that has a certain load rating. At this point, the audit team may request calculations that show the overload capacity of this particular thrust bearing and decide that it does not allow for brief periods of operation in surge. The compressor manufacturer is asked to design the surrounding space such that a larger bearing could be accommodated. The future owners are advised to purchase a next-size-larger spare thrust bearing and update the spare parts list in accordance with this decision.

Phase 3 is an ongoing review of vendor compliance with specifications, monitoring of the vendor's manufacturing progress, and resolution of any issues that could affect timely delivery of a quality product. Phase 3 starts where Phase 2 ends and is winding down after the fully installed machine is handed over to, and accepted by, the owner-operator. This date usually coincides with the completion of test runs at the destination facility.

Detailed Specifications Are Needed

Whenever compressors are required for a project, their duties or performance requirements must be described on a design basis, in a general instruction document, and in one or more specifications. The specification requirement applies to all compressors. For centrifugal process gas compressors, the basic specification may simply be API-617; additional requirements are usually tabulated and referenced by the experienced purchaser. It is not appropriate to simply invoke the somewhat general API-617 as a stand-alone reference document. This is because API-617 contains a number of "bulleted" items which require a decision or choice to be made by the owner-purchaser.[2]

Many of these "bullets" in API-617 refer to options that predefine or delineate extent of supply. As just one example, a decision is to be made on which party will be responsible for base plates common to driver and driven machine. There also are options to be spelled out on control instruments and the like. More detail is often needed: Unless the purchaser defines who will preprime the underside of a base plate and what type of primer is to be used, chances are that this detail will get overlooked. Once the discrepancy is discovered, it will consume much time and money to rectify. At that time there will be pressure to make compromises and long-term reliability will be sacrificed in the interest of expediency and cost-cutting.

The degree of detail given by the specifying entity depends largely on the risk tolerated by an owner-purchaser. Compressor owners must take a position on how much maintenance they are willing to perform and how many planned shutdowns should be scheduled over the projected life of the compressor and its driver. The answers will certainly depend on many factors, including the cost of labor. As mentioned earlier, geographic location or remoteness of the installation and the quality of the workforce members assigned to operational and maintenance-technical tasks are also factors to consider.

Briefing Project Executives on MQA

To be sure of obtaining reliable machinery, the project executive should be made aware of three very important issues. If any of the three are ignored, the reliability of an installation is likely to suffer.

1. Some project budgets allow only for procurement of the least-expensive machine. If one purchases inexpensive compressors, odds are overwhelming against equipment reliability. The machine will likely require more frequent maintenance intervals or will simply fail more often than a compressor that has been subjected to prepurchase MQA, or machinery quality assessment.

2. The project budget must include the cost of an up-front MQA. Together, the up-front audit and subsequent ongoing reliability reviews will amount to approximately 5 percent of the equipment cost.

3. The purchaser's intent to perform part of this machinery quality assessment (MQA) at the bidders' or compressor manufacturer's factory location will cause an expense to the manufacturer. Bidders must be made fully aware of this requirement and their total, all-inclusive cost proposals must reflect this requirement.

Project funding must also reflect the three phases of the machinery quality assessment effort. Phase 1 involves job functions beyond machinery engineers. Three or four experienced specialists participate in Phase 2, the audit. A designated experienced machinery engineer does most of the reliability review work comprising Phase 3 and acts as the focal point person or coordinator of review input contributed by electrical engineers, or welding experts, and so forth. Reliability reviews (Phase 3) involve a machinery engineer full-time; the other subject matter experts contribute on an as-needed, part-time, basis.

Only Competent Manufacturers Are Invited to Submit Bids

Applying wisdom to the compressor procurement process requires that we identify the three or four manufacturers we wish to invite. In other words, we want to devote all of our attention to bidders from whom we would purchase a compressor in the size and flow and pressure range under consideration.

During the early definition phase of a project, a designated individual or team of individuals must compile a list of competent suppliers. Suppliers or bidders must be in a position to represent and support their product in the geographic region where the compressor(s) will ultimately be installed. It must be ascertained that a potential supplier has retained its presumed legacy design and quality control. In many cases it will be necessary to arrange for a plant visit. Qualified compressor vendors are then invited to submit a detailed bid, or proposal.

The question on how long one can run compressors before scheduling turnaround inspection must be answered. Of course, unexpected outages can be extremely expensive and even a scheduled outage event can negatively affect a facility's balance sheet. We have had occasion to listen to opinions as well as field experience on the topic of shutdown frequency from both engineering staff and management personnel. Needless to say, the opinions are divided not only between the two groups, but also among the engineers involved and within the management groups themselves.

With so much money at stake, how should one approach the issue and how might a reasonably precise numerical answer be obtained? Obtaining a very precise numerical answer will forever elude us since there are far too many variables involved.

The life expectancy of centrifugal compressor components is prone to be influenced by gas conditions, purity, corrosivity, and other parameters. Process facilities experience occasional upsets and acceptable control is not always maintained. Operation at varying speeds may subject compressor impellers to different steady-state and alternating stresses. Lube oil quality and purity can easily affect bearings and seal systems. Electronic governing systems for compressor drivers may be influenced by ambient conditions and component drift, or maintenance oversights and operator error. Redundancy may have to be specified for some subsystems. Multiple layers of protection are often highly desirable.

Making a reasonably accurate assessment of prudent operating time to the next scheduled turnaround inspection is difficult. The engineer responsible for it will have to review all of the above and many more factors. A thorough investigation of the experience and service condition of machines similar to those intended for a new project will be of value. Comparisons of stress levels on bearings, seals, impellers or, in the case of axial compressors, identical compressor blades operating elsewhere should be pursued. Evaluation of "their" actual maintenance procedures against "our" projected maintenance procedures would also be appropriate.

We have seen over the years that much of this information is easiest obtained during preprocurement reviews of the particular design offered by a compressor manufacturer. These preprocurement reviews (Phase 1) are not to be confused with design audits (Phase 2) and ongoing reviews (Phase 3). We know of relatively few procurement situations where capable vendors would not show a willingness to explain in detail their prior experience for the service conditions imposed on the proposed machinery. In other words, capability and the ability to demonstrate satisfactory experience go hand in hand. A good vendor has nothing to hide.

Service conditions take into account parameters found in comprehensive data sheets. Most of these parameters are found in the appendix material to API-617. Service condition listings and reviews must include driver startup (slow-roll) time, input power demand, rating, experience relating to gas molecular weight, impeller tip speed, gas temperature, and pressure. Prior experience in these and other parameters should be established and extrapolations from past designs pointed out. Along these lines, capable vendors can usually demonstrate mechanical design experience for applicable critical parameters which could include:

- Bearing span
- Bearing design, loading, size, and clearance
- Impeller design—structural and thermodynamic
- Casing size and design
- Nozzle or guide vane orientation
- Casing-joint design configuration, gasket types selected, and bolting
- Coupling design and arrangement
- Surge limits and proposed surge control schemes
- Material selection
- Compressor suction and discharge nozzle sizing for conservative gas velocity
- Compressor discharge nozzle sizing for acceptably low gas velocity under future upgrade (especially, increased throughput) conditions
- Number of stages and staging arrangement
- Power transmission components (design and arrangement of gears, couplings, and coupling guards)
- Rotor dynamics
- Sealing systems

To then lay the groundwork for answering future questions on mean-time-between-turnaround inspections, the responsible engineer (the owner's engineer) would do well to expand his or her review of a vendor's mechanical design experience so as to include requests for operating and maintenance feedback from other installations.

During the second phase of MQA, the review engineer makes use of a checklist. He or she will ascertain compliance with specifications and will ask for whatever data are needed to ascertain a machine reliable.[3] Among the checklist items will be

- Completed API data sheets
- Cross-section drawing with internal radial and axial clearances
- Labyrinth clearances
- Dimensioned general arrangement drawing showing all process gas, cooling water, and electrical connections
- Input and results of rotor sensitivity (dynamics) study, including bearing, support, stiffness, and damping behavior as a function of rotor speed
- Input and results of a torsional critical speed analysis, if required
- Individual impeller performance curves indicating polytropic head and impeller efficiency
- Overall compressor performance curves
- Dimensioned thrust-bearing drawing and final calculated thrust load
- Dimensioned radial bearing drawings and loadings
- Driver versus load torque/speed demand curve
- Dimensioned coupling cross-section drawings with shaft end assembly procedures and fits
- Input data for a rotor thrust analysis by third party
- Results of vendor's axial load (thrust) analysis with both new and worn balance piston labyrinth clearances
- Dimensioned shaft seal drawing
- Alignment offset data (to accommodate thermal rise)
- Procedures for factory and field test runs
- Minimum stable flow
- Discharge temperature
- Design, settling-out, and hydrostatic pressures
- Flange size, rating, face, orientation
- Nozzle velocity
- Impeller gas velocity, high MWs (limiting allowable tip speed)
- Thrust bearing type, capacity, load, velocity
- Impeller construction
- Impeller keys
- Impeller stress and yield point (H_2S)
- Impeller widths and diaphragms
- Impeller and sleeve material, H_2S-tolerant

- Impeller overspeed test data
- Casing materials and impact test data
- Piping material and limit of supply
- Spare rotor and gears
- Coupling attachment (hydraulic) and tools for installation and removal
- Lube data and compatibility
- Lubricant pour point, inner seal
- Coupling (adapter plate, spacer, limited end float)
- Base plate versus soleplate selection criteria
- Driver sizing
- Compressor control details
- Responsibility for torsional critical speed calculation
- Lateral vibration monitoring parameters and limits
- Design pressure for auxiliary and support equipment
- Noise data
- Shop tests (mechanical, gas loop, noise, consoles, driver, test stand capability, seals)
- Auxiliary system schematics and bill of materials (see seal and lube oil systems)
- Auxiliary piping material and layout
- Outline drawings for auxiliary systems
- Nozzle size, orientation, facing geometry, bolting sizes, and torque requirements
- Maintenance weights
- Vents and drains
- Fixed and flexible supports
- Shaft movement, hot versus cold X-Y-Z movements
- Shaft taper details
- Lifting dowels
- Jacking screws (positioning and alignment brackets)
- Support shims
- Grouting details and procedures
- Coupling axial freedom
- Variable frequency control on motors
- Rotation agreement for driver and compressor
- Section drawings

- Coupling installation and removal procedures
- Coupling dial indicator access
- Thrust-bearing oil supply and drainage
- Thrust-bearing type, size, etc.
- Thrust-bearing temperature monitoring
- Radial bearing stabilization (anti-oil whip)
- Vent space for bearings and seals
- Inner seal oil pour point compliance
- Seal maintainability
- Seal bypass leakage vulnerability and proposed means of monitoring
- Casing horizontal joint bolts close to seal
- Seal O-ring material
- Seal balance line and chamber
- Seal radial clearance
- Shaft step windage
- Balance piston and clearance details
- Balance chamber bleed capacity
- Balance line size
- Balance line flushing
- Labyrinth material
- Labyrinth maintainability
- Sleeves under seals and labyrinths
- Sleeves sealing at shaft
- Impeller width
- Impeller locking provisions
- Impeller vane erosion vulnerability
- Flushing nozzle entry
- Flushing nozzle strainer
- Flushing nozzle on-stream maintainability
- Side entry locations
- Side entry wheel space
- In-out casings (contamination)
- Divider diaphragm
- Horizontal joint seal
- Vibration probe location

- Vibration probe mechanical support
- Coupling guard venting and strength of housing
- Coupling balance and flexibility
- Seal configuration and safety backup
- Drain location and absence of windage
- Vibration analysis results
- Acceptable location of vibration probes
- Flexible supports
- Coupling weight
- Correlation with measured critical speeds
- Cross-check of calculations

Contractor Piping and Instrumentation

Inlet System

- PI (pressure indicator) and TI (temperature indicator) supplied
- Heat tracing
- Flare release valve
- Minimum pressure protection
- Knockout drum LHCO (level-high-cut-off)
- Liquid drainage collecting drum and casing drains
- Motorized valve with local VPI (valve position indicator) and switch
- Minimum temperature protection (deicing at air inlet)
- Local controller for butterfly valve
- Butterfly operation and linkage good for high vacuum conditions
- For high pressure, remote block valve bleeder bypass
- Recycle entry upstream of knockout drum
- Liquid injection for refrigeration recycle
- Blind for gas

Discharge System

- Check valve protects recycle
- Safety vent inside blocks valves
- Blind

- Drain
- Flow meter sees net flow
- Recycle (vent) controller sees gross flow
- Recycle and flow control at local panel
- Recycle vent valve can pass 70% flow at all molecular weights
- Verify that vent line pressure drops are not limiting
- Silencer on vent

Foundation and Compressor House Layout

- Foundation
- Concrete under all base plate members
- Not shared with reciprocating compressors
- Height set by suction line and oil drain slope
- House layout
- Floor at top of base plate, not foundation

Accessibility and visibility of panel location

- Machine visibility from panel
- Crane reach
- Crane capacity
- Dropout area
- Floor space for maintenance
- Overhead seal tanks visible
- Components inside oil reservoir, if any
- Oil vent away from air inlet
- Steam vent away from air inlet
- Clearance for pulling pumps, filters, coolers
- Electrical classification of panel

Process piping layout

- Heat tracing and insulation
- Liquid drains
- Straight runs at fluid machinery inlets
- Turning vanes in pipe elbows
- Strainer access and strength

- Unduly close location to other components
- Basket mesh, strength
- Cleaning access
- X-Y-Z stops close to optimum locations
- Adjustability provisions for each
- Ruggedness
- Silencer
- X, Y, or Z stops at first elbows
- Located to channel thermal growth away from machine
- Stop torques and moments
- Access for internal inspection of large pipe supports
- No cold springing
- Avoid spring supports
- Temporary support provisions at relief valves and blinds
- Suction drum crinkled wire mesh screen material and support
- Atmospheric inlet screen at compressors, if needed
- Provision for air run-in, if needed
- Discharge pipe review
- Adjustable X-Y-Z stops at discharge pipe
- Drain provisions and vacuum breaker vents
- Check valve damper actuation and access
- Silencer for pipe blow-out

Fabrication Erection and Cleaning Procedure

- Silencer sectional drawing
- Spare parts definition
- Manufacturer's representative identified and arranged for
- Erection coverage defined
- Startup coverage defined

What We Have Learned

Machinery quality assessment is an up-front activity that typically consumes 5% of the cost of machinery. This cost must be allocated and reflected in the project budget.

Without performing MQA on process gas compressors it will be near impossible to build a highly reliable plant. New projects that incorporate the least expensive first cost "bare-bones" compression equipment and appurtenances will not achieve the reliability and availability of plants that have made it their business to purchase optimized machinery.

MQA identifies optimized machinery. The cost of MQA is usually retrieved by the time a facility starts up. It has been estimated that the MQA effort pays back 10:1 over the projected life of a unit or plant.

MQA looks at more than just the compressors. Piping, layouts, auxiliaries, and vessels are among the items reviewed.

Checklists are extremely helpful; many are found in Ref. 3. Obtaining answers to checklist items represents a networking, training, mentoring, and educational opportunity.

References

1. Bloch, Heinz P., *Improving Machinery Reliability*, 3d ed, Gulf Publishing Company, Houston, TX, 1982. Revised 2d and 3d ed.
2. American Petroleum Institute, API Standard 617, Centrifugal Compressors, Alexandria, VA.
3. Bloch, Heinz P., and Fred Geitner, *Maximizing Machinery Uptime*, Elsevier-Butterworth-Heinemann, Stoneham, MA, 2006.

CHAPTER 11

Compressor Failure Analysis Overview

To demonstrate how an effective failure analysis should be framed and carried out, we opted to use a compressor-related example. This case history involves a three-dimensional impeller. It demonstrates the straightforward thinking processes and analysis approaches that have worked well for us in the past.

It has been said that failure analysis and troubleshooting must be preceded by stating what the problem is. The problem is always a deviation from the expected norm. In the overwhelming majority of cases, more than one seemingly minor deviation combines with other seemingly insignificant deviations. The combined minor deviations then lead to the observed problem. Not addressing compressor problems can lead to calamities that range from simple annoyance to utter devastation.

Problems cannot be solved without data. Repairs carried out without understanding what led up to a failure will usually result in repeat failures. So, we must first gather data and then organize the data into evidence of things good and evidence of things bad or defective. We also need timelines and time-based observations. We must pay attention to actions and events that preceded the failure. All actions and events are of interest, the acceptable as well as the questionable, the interesting ones as well as those that might have bored you when you first found out about them.[1]

From an interrogation of timelines and time-linked observations we can ask "So what?" Answers to our questions allow us to make a good guess, or a best guess. Occasionally, we benefit from additional data—data needed to justify our best guess. The data must at all times line up with science and every scientific law on the books. Speculations to the contrary are unprofessional and are a waste of time. To assume that our company is the first ever to experience a particular type of failure is not something we encourage. Similarly, to think that ours will become the first commercial enterprise that will get away with numerous deviations occurring simultaneously is an unprofitable assumption that reliability professionals should not endorse.[2]

Case History Dealing with Impeller Failure

Suppose we had to determine why a three-dimensional impeller, similar to the style shown earlier in Fig. 7.2 or as depicted in Fig. 11.1, had failed. Suppose further that this "mixed-flow" impeller incorporates a number of freestanding blades and the blades are attached by welding. A three-step approach allows competent reliability professionals to quickly zero in on the most probable root causes of failure.[3] The analysis starts out by explaining the seven root cause methodology, the "FRETT" examination and, finally, appropriate model validation procedures.

Failure Analysis Step No. 1: The "Seven Root Cause Category" Approach[4]

The first of the three failure analysis steps is called the "Seven Root Cause Category Examination." It is solidly based on the premise that all machinery failures fall into one or more of only seven possible cause categories:

1. Design errors
2. Material defects
3. Fabrication and processing errors
4. Assembly and installation deficiencies
5. Maintenance-related or procedural errors

FIGURE 11.1   A single-piece three-dimensional compressor impeller.

6. Unintended operating conditions

7. Operator error

We use a process of elimination to determine which of the seven cause categories should be deleted and which one, or perhaps two, hold the key to a failure event. Using logical thought processes, we narrow things down. First off, we would simply ask ourselves which of these seven cause categories are influenced by the compressor user, and which ones are under the full jurisdiction of the compressor manufacturer. The answer determines the cause categories where failure analysis efforts should be concentrated.

We must start with what we know about this compressor. We must have data; we cannot solve a problem without data. Therefore, we must examine what the record shows and what the record does not show.

Suppose the impeller had 19 blades and there is no indication in the record that its user/owner had ever operated the machine at unintended speeds or other unauthorized conditions. Also, all gas properties had been disclosed to the vendor at the inception of the procurement chain. Had there been prolonged surge (operation at low flow), we would have expected thrust bearing damage, but no such damage had been experienced at the time of blade failure.

We assume it had thus been ascertained that no operator error occurred. Accordingly, categories (6) and (7) are now being ruled out. Assume further that during compressor maintenance there is no logical causal event fitting into the assembly and installation, or maintenance/ procedural error categories. Anyone making the statement "it could have been an assembly error" would have to explain what exactly could have been misassembled and would have to show data or measurements in support of such a claim or statement.

With (4) and (5) thus being ruled out, we might focus next on item (2), material defects, but again find no metallurgical evidence of flaws in the base material selected by the manufacturer.

The reviewer would thus be left with the two cause categories design error (1) and fabrication and processing errors (3). We would be left with these two for closer investigation and scrutiny.

Failure Analysis Step No. 2: "FRETT"[4]

It is universally recognized that all machinery component failures, or machine distress brought down to the mechanical component level, are attributable to one or two of only four failure mode sets. These four possibilities are force, reactive environment, time, and/or temperature. An easily remembered acronym, "FRETT," allows us to recall these four possible initiators. FRETT is such an important concept that we will repeat the message by formulating it in different words:

All machine components will fail due to deviations in allowable force (F), or exposure to a reactive environment (RE), or time (T) in one extreme or another, or temperature deviation (T). As we look at a component and study its operating history, we should be able to determine what deviation was involved.

Getting back to our impeller example, assume we had studied the matter. Studying includes reviewing accumulated data from various sources and ascertaining some facts. Say, hundreds of compressor impellers in identical gas service had accrued operating hours far in excess of those at issue in this failure event. Hence, the impeller did not fail because the machine ran for too long and we could immediately rule out "time."

One would also rule out "temperature," since (assume you were able to ascertain) the actual compressor-operating temperatures had always remained well within the acceptable range. Likewise, one might discount the suspicion that corrosion (a reactive environment) was responsible for the failures if (a) it could be shown that the owner's gas composition did not measurably deviate from that disclosed in the original specification documents and (b) a metallurgical exam showed no evidence of such corrosion.

That clearly would leave the reviewer with "force" as the only remaining logical failure mode set. The issue might be what type of force acted on the impellers, or where the force came from. This is where one might rely on data collected by suitable instrumentation and data collector modules. Remember, the failed impellers had 19 blades and it is well known that failures can occur due to blade passing frequencies (BPFs) coinciding with impeller natural frequencies. We would recall the discussion on vibration behavior (Chapter 5) and look for amplitude excursions at a frequency of 19-times-RPM, 38-times-RPM, and other multiples. Moreover, we would search for the presence of acoustic pulsing at BPF in the balance line connecting compressor discharge and first-stage suction inlet. Unless the owner has in-house expertise, a consulting company would be selected to do further investigation. They would probably find blade-passing frequencies (or multiples) that coincide with the impeller speed. Such coincidences are causing resonant vibration and fatigue failure at the welded junction of blade and hub.

Or, we would recall that, in some services, there is compressor fouling (Figs. 11.2 and 11.3). Chunks of hard or solid deposits may come loose and strike the downstream blading while the compressor is operating. The two arrows in Fig. 11.2 indicate that foulants can become rather thick; continuous on-stream injecting of up to 3 percent (by weight) of a suitable liquid is recommended. Part of the liquid must be injected into the suction piping and the remainder gets injected through small nozzles into the diaphragm passages. Intermittent liquid injection is discouraged because it can be ineffective. Continuous injection of flush liquid into the compressor will prevent formation and deposition of foulant, as will certain coatings applied to compressor impellers.

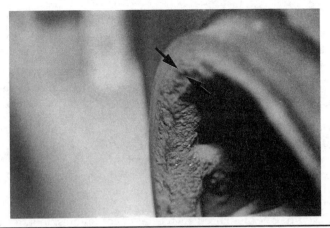

FIGURE 11.2 Fouling deposits in the eye of a closed impeller (Ref. 5).

FIGURE 11.3 Fouled compressor diaphragm passages.

If solid foulants are present, we might pursue *"Force"* as the potential basic agent of part and component failure mechanisms. Depending on the distance traveled by a solid chunk of fouling deposit, the force of impact could be calculated and an assessment made as to the overload this might have represented on a blade.

Step No. 3: Validation, or Relating Analytical Models to Field Experience

Whenever an impeller design or other component is "modeled" for computer analysis, the analyst will generally make a number of assumptions and the validity of these assumptions must be ascertained. Common sense tells us that there are four possibilities and corresponding definite conclusions, as follows:

1. Analysis says "OK," but compressor part fails: model or analysis technique flawed

2. Analysis says "OK," and compressor part survives: model and analysis technique OK

3. Analysis says "not OK," compressor part also fails: model and analysis technique OK

4. Analysis says "not OK," but compressor part survives: model or analysis technique flawed

But suppose the manufacturer's analysis claims that the design of a failed component was acceptable, yet the owner experienced repeat failures. In the absence of root cause factors other than design and/or fabrication, the reviewer would be led to the logical conclusion that the manufacturer's model or analysis technique was flawed.

If you have experienced such component failures, look for tacit acknowledgement of this conclusion on the part of the vendor. In the case of impeller failure, the manufacturer perhaps no longer recommends operating these compressors at speeds that correspond to the BPF interference mentioned earlier. Or, the manufacturer switched from welded blades to a superior fabrication method employing four-axis machining on a single block of material of suitable metallurgical composition. That would support the notion of a flawed computer model. You would concentrate your investigative work on the success of milled impeller redesigns. These redesigns could logically be expected to cure the problem.

Whether we are dealing with a compressor component failure event or an unexplained compressor wreck, there are four cardinal rules that will apply:

1. Make sure things don't get worse if you can help it. Get others involved. "Others" are knowledgeable contributors and might include a plant superintendent, unit supervisors, and the fire department. Appropriate investigative techniques might include hazardous operations (HAZOP) analysis and consulting the rather straightforward troubleshooting tables reproduced elsewhere in this text (see index).

   The immediate involvement of other personnel is especially valuable if an immediate emergency shutdown is required (true in the event of major hydrocarbon leaks, flying parts, fires, toxic gas releases, major safety or environmental issues, very loud unusual noises, or vibration excursions that keep getting worse).

2. Make sure shut down data are saved that could help in future troubleshooting (for example: vibration levels, pressures, temperatures, process conditions).

3. Gather data. Get all data logs, computer data, vibration data, pressures, temperatures, and flows. Debrief or talk with

operators, process technicians, mechanics, and others on that shift, determine repairs and what was done, process changes, etc. Determine what parameters have changed in the time period between good operation and bad operation. The machine was operating OK a few days ago and now it clearly is not. What has changed? Get experts and equipment manufacturer field service specialists to help review data.

4. Make a decision based on available data and ask which parts are probably affected. Then determine if a shutdown is needed now, or if the machine can be safely operated prior to a scheduled down time. Develop plans for either case.

Troubles, causes, and remedies are collected and displayed in chart form. For each event, there may be one or two causes, at most. Interrogate your own data and eliminate or delete causes for which you have no evidence. Again, each can be considered and deleted if there are no data supporting a potential cause. Use a process of elimination.

Trouble	Possible Cause	Remedy
Low lube oil pressure	Oil leakage	Tighten flanged or threaded connections. If necessary shut down compressor and replace defective gaskets.
	Faulty lube oil pressure gauge or switch	Calibrate or replace the faulty instrument.
	Low oil level in reservoir	Add oil to bring the reservoir to the proper operating level.
	Oil pump suction plugged	Secure the compressor, drain lube oil reservoir. Inspect and clear pump suction.
	Leak in oil pump suction piping	Tighten leaking connections or shut down compressor, drain lube oil reservoir, and replace gaskets.
	Clogged oil strainer or filter	Inspect and clean oil strainer or filter.
	Failure of both main and auxiliary oil pumps	Check the operation of both pumps according to manufacturer's instructions.
	Relief valve set too low or stuck open	Check relief valve for correct setting and/or for proper operation.
	Incorrect pressure control valve setting	Check control valve for correct setting.

Trouble	Possible Cause	Remedy
Excessive bearing oil drain temperature	Inadequate or restricted flow of lube oil to bearings	Check oil pressure gauge. If pressure is below design, see "Low Lube Oil Pressure" in "TROUBLE" column. If pressure is satisfactory, check for restrictions in flow of lube oil to the affected bearings; the orifice plug, lube oil strainer, or filter may be plugged. Check sight flow indicators (if provided) for proper oil return flow to the reservoir.
	Poor condition of lube oil or dirt or gummy deposits in bearing resulting from contamination.	Minimize contamination of oil by increasing frequency of oil changes until condition is cleared up. Inspect lube oil strainer or filter—clean more frequently, if necessary. Inspect bearing for cleanliness.
	Inadequate cooling water flow through lube oil cooler	Increase cooling water flow/pressure through the lube oil cooler. Check for above design cooling water inlet temperature.
	Fouled lube oil cooler	Inspect lube oil cooler—clean, if necessary.
	Wiped bearing	Inspect journal bearings and replace as required. Determine cause of bearing failure.
	Excessive oil viscosity	Check required oil viscosity w/MTS section and select lowest permissible.
		NOTE: Lube oil temperature leaving the bearings should never exceed 180°F.
Excessive bearing wear	Shaft misalignment; rough journal surface	Correct shaft alignment. If scoring is excessive, replace shaft and recondition damaged journal.
	Vibration	See "Excessive Vibration" in "TROUBLE" column.
	Lack of proper lubrication	Check for dirt or obstructions in oil feed lines and orifices.
	Dirty oil	Check condition of lube oil. Recondition or replace oil as required. Check filters to ensure they did not collapse.

Compressor Troubleshooting Guide (*Continued*)

Trouble	Possible Cause	Remedy
Lube oil temperature entering bearings too low	Excessive cooling water flow through lube oil cooler	Decrease cooling water flow through lube oil cooler, check water inlet temperature, and confirm water control valve is properly sized for adequate flow control.
Water in lube oil	Leak in lube oil cooler tube or tubes	Hydrostatically test the tubes and tube sheet for leaks and repair as necessary.
	Condensation in oil reservoir	During operation, maintain a minimum lube oil reservoir temperature of 120°F to permit separation of entrained water. To avoid excessive oil vaporization do not exceed a reservoir temperature of 140°F.
Loss of compressor discharge pressure	Excessive compressor inlet temperature	Investigate and correct the cause of the excessive inlet temperature.
	Leak in discharge piping	Inspect discharge piping for escaping gas. If a leak exists, shut down the compressor and repair.
	Excessive system demand from compressor	Check operation of all control valves, including those in recycle loop.
Compressor surge	Inadequate flow through compressor	Refer to performance curves.
	Change in system resistance due to obstruction or improper valve position	Check the position of all valves and for faulty control valve (if installed).
	Deposit buildup on rotor, or diffuser, restricting gas flow	Employ available liquid flushing techniques. If results are not satisfactory, mechanically clean the impellers.
Excessive vibration	Improperly assembled parts	Listen for rubbing noises, particularly seals. Locate source of noise, shut down the compressor, dismantle, and inspect. Repair as necessary.
	Loose or broken bolting	Check bolting at casing-bearing case supports. Check soleplate bolting. Tighten or replace as necessary.

Compressor Troubleshooting Guide (*Continued*)

Trouble	Possible Cause	Remedy
Excessive vibration (*Continued*)	Piping strain	Inlet and discharge piping should have been installed in such a manner so as to limit piping strains. Inspect for proper installation of pipe hangers, springs, or expansion joints. Examine the piping arrangement and correct, as necessary.
	Sympathetic vibration	Adjacent machinery can cause vibration of the compressor even when shut down, or at certain speeds due to foundation or piping resonance. A detailed investigation is required in order to take corrective measures.
	Fouled rotor	Shut down the compressor and inspect the rotor. Liquid flush or abrasive clean, as necessary.
	Coupling misalignment	Check coupling alignment with compressor at operating temperature. If misalignment exists, correct.
	Failure of coupling elements	Replace coupling.
	Damaged rotor	Shut down the compressor and inspect the rotor. If damaged, replacement and rebalancing of rotor is recommended.
	Excessive bearing clearance	Shut down the compressor and check bearing clearances. Check for rough or uneven surfaces and other evidence of pounding. If clearances are excessive, replace the bearings.
	Liquid "slugs" striking impeller	Locate source of liquid and correct. Drain compressor casing of any accumulated liquid.
	Bent shaft	Check shaft runout, if excessive, recondition or replace the shaft.
	Shaft and impeller not balanced as an assembly	Rebalance shaft and impeller assembly.

Compressor Troubleshooting Guide (*Continued*)

What We Have Learned

Buying reliable equipment is the most cost-effective path to failure avoidance. If one wants to have reliable compressors one must expend up-front effort. This preinvestment was described in Chapter 10. Called Machinery Quality Assessment (MQA), it will pay handsome dividends for decades and is the best path to failure avoidance.

When something does fails, we must use a structured approach and dig for a deviation buried in our own data. Without data, we are at best taking guesses and are setting ourselves up for repeat failures.

There are only seven failure cause categories; they are listed in the table. Paying attention to these categories, we must go through a process of elimination to find the one, or at most two, wherein the flaw or defect is rooted.

We can be certain that compressor parts can only fail if a limiting value of F, RE, T, or T is exceeded. We know that deviations or violations tend to combine and cause major problems.

When doing failure analysis and troubleshooting, all findings and conclusions must be supported by data and must conform to science. Because the failure of compressor components is not influenced by the supernatural, there will be logical explanations for everything.

Every effect has a cause, and in nearly 100% of all failure incidents a series of small contributing causes successively reduced the component failure margin. Ultimately several—in themselves small—deviations will combine into one big problem.

A conscientious machinery engineer will act even on small deviations.

References

1. Bloch, Heinz P., *Practical Guide to Compressor Technology*, 2d ed, John Wiley & Sons, Hoboken, NJ, 2006 (1st ed also available in Spanish).
2. Bloch, Heinz P., *Improving Machinery Reliability*, 3d ed, Gulf Publishing Company, Houston, TX, 1998.
3. Bloch, Heinz P., and Fred Geitner, *Machinery Failure Analysis and Troubleshooting*, 4th ed, Elsevier Publishing, New York, London, Tokyo, 2012.
4. Bloch, Heinz P., and Fred Geitner, *Major Process Equipment Maintenance and Repair*, 2d ed, Gulf Publishing Company, Houston, TX, 1997.
5. Hanks, Julian, Inter-company report on equipment revamping and upgrade opportunities, 2002, unpublished.
6. Bloch, Heinz P., and Fred Geitner, *Maximizing Machinery Uptime*, Vol. 5. *Series on Practical Machinery Management*, Gulf Publishing Company, Houston, TX, 2006.
7. Bloch, Heinz P., and J.J. Hoefner, *Reciprocating Compressors: Operation and Maintenance*. Gulf Publishing Company, Houston, TX, 1996.

Bibliography

Bloch, Heinz P., and Claire Soares, *Process Plant Machinery*, 2d ed, Butterworth-Heinemann, Stoneham, MA, 1998.

Bloch, Heinz P., and Abdus Shamim, *Oil Mist Handbook: Practical Applications,* Fairmont Publishing, Lilburn, GA, 1998.

Bloch, Heinz P and Claire Soares, *Turboexpanders and Process Applications,* Gulf Publishing Company, Houston, TX, 2001.

Bloch, Heinz P., and A. Godse, *Compressors and Modern Process Applications,* Wiley & Sons, Hoboken, NJ, 2006.

Bloch, Heinz P., and Dr. M. Singh, *Practical Guide to Steam Turbine Technology,* 2d ed, McGraw-Hill, New York, 2008. (1st ed, 1995, presently available also in Spanish.)

Bloch, Heinz P., and Fred Geitner, *Introduction to Machinery Reliability Assessment,* 2d ed, Gulf Publishing Company, Houston, TX, 1994.

Bloch, Heinz P., *Practical Lubrication for Industrial Facilities,* 2d ed, Fairmont Publishing, Lilburn, GA, 2009.

Bloch, Heinz P., *Compressors and Expanders: Selection and Applications & Contributors: for the Process Industries,* 1982. Marcel Dekker, New York.

Bloch, Heinz P., and Fred Geitner, *Machinery Component Maintenance and Repair,* 3d ed, Gulf Publishing Company, Houston, TX, 2004.

Bloch, Heinz P., and Allan Budris, *Pump User's Handbook: Life Extension,* 3d ed, Fairmont Publishing, Lilburn, GA, 2006/2008/2011.

Bloch, Heinz P., *Pump Wisdom: Problem Solving for Operators and Specialists,* John Wiley & Sons, Hoboken, NJ, 2011.

CHAPTER 12

Reciprocating Compressors: Background and Overview

Introduction

Reciprocating compressors are flexible, efficient, have a wide range of applications, and can generate high heads or discharge pressures independent of gas density. They are the most common type of compressors. Worldwide, the total installed horsepower of reciprocating compressors is about 2½ times that of centrifugal compressors.

The history of the reciprocating compressor dates back to antiquity when bellows, wooden cylinders, and pistons were used to produce air blasts. The forerunners of the modern compressor did not make any noticeable progress until the middle of the 19th century. With industry making rapid advancements, particularly in the United States, a need developed for transportable energy in the form of compressed air. Compressed air conducted in pipes was considered the best method, and somewhat superior to steam.[1]

Compressed air found its first practical large-scale application in the construction of tunnels. In the United States, the first such machine was used when the Hoosac Tunnel was constructed in 1866. The machine was a four cylinder, single-acting horizontal unit. It had poppet valves, used water for cooling, and was driven by a turbine wheel.

The compressor was later taken to a marble quarry where compressed air for power drilling rapidly became an established practice. With the rapid progress of industry, new processes came into being and many new uses were devised for compressors. The petroleum industry took the lead and a large percentage of hydrocarbon-processing applications are still handled by reciprocating compressors.

Basic Operating Principle Explained

A reciprocating compressor raises the pressure of a gas by forcing a volume reduction. It accomplishes this through the movement of a piston, which is to say by the displacement of gas in a cylinder. In essence, reciprocating compressors are fixed capacity machines. Variable capacity may be obtained either by changing the speed of the prime mover, by the use of suction valve lifters, or by the application of clearance pockets. Through the use of unloaders, it is possible to maintain constant horsepower over a wide range of pressure conditions, thus enabling economic operation of the driver at all times.

Reciprocating compressors are designed and manufactured as both air-cooled (see Fig. 12.1) and water-cooled models (Fig. 12.2). Power ratings vary from fractional to more than 40,000 HP (30,000 kW) per unit. Pressures range from low vacuum at suction, to 30,000 psi (2000 bar) and higher at discharge for special process compressors. Small air compressors are usually single acting whereas the majority of the higher power machines are double acting. Double-acting compressors have pistons that compress gas at both ends of the

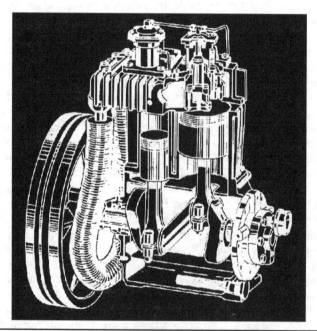

Figure 12.1   Small air-cooled, single acting two-stage reciprocating compressor.

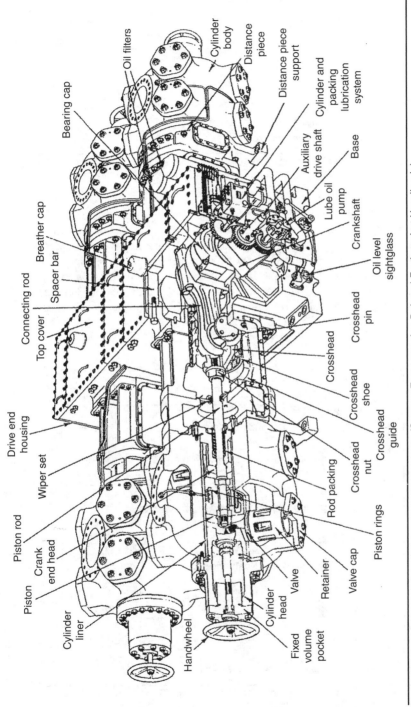

FIGURE 12.2 Large reciprocating process compressor. (*Source: Transamerica DeLaval Engineering Handbook.*)

143

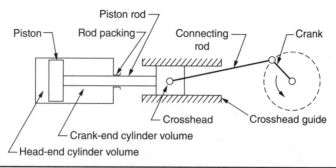

Figure 12.3 Double-acting design principle—suction and discharge valves are not shown. (*Source: Transamerica DeLaval Engineering Handbook.*)

compressor cylinder. While the piston is on its discharge stroke on one end, the same piston is on its suction stroke on the other end of the same cylinder (see Fig. 12.3). Their strength lies in low-to-moderate flow (600–3000 ft³/min or 1000–5100 m³/h) applications, especially with low-molecular-weight gases, such as hydrogen and in applications where high discharge pressures (2000–5000 psi or 138–345 bar) are required. A map of reciprocating compressor capacity versus pressure capability is shown in Fig. 12.4.

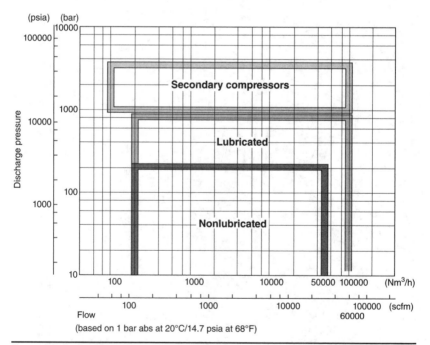

Figure 12.4 Reciprocating compressor application map. (*Source:* Burckhardt Compression, Winterthur, Switzerland.)

Reciprocating compressors are furnished with either a single stage or with many stages of compression (multistage). The number of stages is determined by the overall compression ratio. Compression ratio per stage and valve life are generally limited by the discharge temperature and the practical ratio usually does not exceed 4. However, small-sized units in intermittent duty are often furnished with compression ratios as high as 8.

On multistage machines, intercoolers are normally provided between stages. Intercoolers are heat exchangers which remove the heat of compression from the gas and reduce its temperature to approximately the temperature existing at compressor suction. At some constant pressure, a given mass of gas at low temperature takes up less volume than the same mass at a higher temperature. Therefore, cooling reduces the actual volume of gas going to the high-pressure cylinders, decreases the horsepower required for compression, and maintains the discharge temperature within safe operating limits.

Reciprocating compressors should be supplied with clean gas because they cannot satisfactorily handle liquids and solid particles that may be entrained in some gas supplies. Liquids and solid particles tend to degrade cylinder lubrication, damage valves, and cause cylinder liners to wear. Facilities such as filters, separators, and liquid knock-out drums upstream of the compressor are nearly always required.

Gas cylinders are generally lubricated, although a nonlubricated design is available when specified. Services usually requiring oil-free compression are, for example, nitrogen, oxygen, and instrument air services.

Most process-reciprocating compressors are designed to one of two prevalent industry-standard specifications: API Standard 618, *Reciprocating Compressors for Petroleum, Chemical, and Gas Industry Services*[2] and API Specification 11P *Specification for Packaged Reciprocating Compressors for Oil & Gas Production Services.*[3]

The API 618 standard covers low-to-moderate speed compressors, typically in the 300 to 700 RPM speed range. These compressors may be of a separable balanced opposed design as shown in Fig. 12.2, but an "integral" design is also fairly common. An integral design refers to a compressor driven by a gas engine where the power cylinders of the engine that turns the crankshaft are in the same housing as the gas compression cylinders—see Figs. 12.5 and 12.6.

In standard API 618 compressors, the driver is separate from the compressor. These machines have historically been used in refineries and chemical plants, or the "downstream" section of compressor applications. Integral compressors use power cylinders whose pistons apply rotational torque to the same crankshaft that imparts motion to the compressor pistons. Integral compressors are often applied in pipeline service and also in inlet compression service at field gas plants. With the passage of time, the trend has been to replace old-style

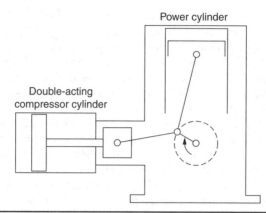

Power cylinder

Double-acting
compressor cylinder

Figure 12.5 Principle of an integral engine compressor. (*Source: Transamerica DeLaval Engineering Handbook.*)

Figure 12.6 Integral engine compressor. (*Source:* Knox County Museum—Mt. Vernon, Ohio.)

slow-speed integrals with gas turbine–driven centrifugal pipeline compressors, packaged medium-speed, gas engine–driven integrals, or electric motor–driven high-speed reciprocating compressors.

Separable API 618 compressors are relatively expensive. One way to reduce their overall cost in refining or other downstream applications is to group individual compression cylinders for multiple services. That means placing one or more cylinders for each service on the same frame and drive motor, an approach known as multiservicing. Careful engineering is needed in such instances and overload capacity needs to be built into the frame. For conservatism and low failure risk, operating loads on piston rods should not exceed 85 to 90 percent of the manufacturer's maximum, continuous, combined rod load rating at relief valve setting.

Reciprocating compressors for hydrogen recovery in the PSA (Pressure Swing Adsorption) process serve as a good illustration of multiservicing. Here, one of the services consists of product-hydrogen compression and another of compressing the feedstock to the PSA unit. In addition, typically, the tail-gas or off-gas is compressed before being sent to a fuel header.

A typical multiservice hydrogen PSA compressor has two cylinders on feed compression, two cylinders on tail-gas compression, and one or two cylinders on hydrogen compression. Instead of three separate reciprocating compressors, only one multiple-throw unit is required. Compared with other layouts or geometries, capital and installation costs are lower with this configuration.

Reciprocating compressors excel over centrifugal compressors in that their design is very tolerant of molecular weight variations. This is of value in the hydrogen-recovery PSA process wherein the tail gas drawn from the PSA adsorption vessel varies in molecular weight. A centrifugal compressor in the same application would be subject to surge or flow reversal, unless it either employed variable-speed control or was designed to accommodate the variable-head requirements while retaining a fixed-speed design.

The API 11P standard covers packaged high-speed separable compressors with speeds from 600 to 1800 RPM. High-speed separables are frequently used in upstream applications such as field natural gas gathering and compression, as well as midstream gas transmission pipeline and gas plant duties. A useful overview and comparison can be obtained from Table 12.1:

	API 11P	API 618
Power, HP (kW)	35–5000 (25–3750)	1,500–35,000 (1,100–26,000)
Speed (RPM)	600–1800	250–600
Footprint and weight	Smaller	Larger
Efficiency	Moderate	High
Operating life	Shorter	Longer
Purchase cost	Lower	Higher
Installation cost	Lower	Higher
Maintenance cost	Higher	Lower

TABLE **12.1**   Comparison of API 11P and API 618 Compressors

A further differentiation may be made by looking at average piston speeds. Average piston speed is the average linear speed at which the piston travels during one revolution of the crankshaft, or through two stroke lengths. It is defined as follows in Eq. 12.1:

$$\text{Average piston speed} = \frac{2 \times \text{stroke} \times \text{RPM}}{C} \text{ ft/min (m/s)} \quad (12.1)$$

where stroke = in (mm)
    C = 12 (60,000)

Average piston speed is an important parameter because it influences compressor MTBF or MTBR. Higher piston speeds usually result in accelerated wear of a reciprocating compressor's sliding and rubbing interfaces. High-speed separables require more frequent and, possibly, more extensive maintenance than their more conservatively designed low-speed counterparts. To expect otherwise can lead to disappointment.

However, Ref. 4 provides an interesting addition to Table 12.1 by alerting owner-purchasers to the existence of a third category of reciprocating compressors, namely the moderate-speed, short-stroke compressor line. This is shown in Table 12.2.

Today's moderate-speed, short stroke compressor line finds use in typical upstream operation. In general and as of 2012, these reciprocating compressors are driven by a natural gas–fueled reciprocating engine. Made of lighter components and produced in relatively high volumes, these compressors are less expensive than the slow-speed, reciprocating compressors that meet the API 618 standard. Their smaller, lightweight frames and low-mass moving components relate to smaller reciprocating forces and moments. They are often skid mounted or "packaged," which reduces installation costs considerably. Packaging results in a self-contained module. No utilities—such

	Low Speed (Long Stroke)	Moderate Speed (Short Stroke)	High Speed (Short Stroke)
Stroke, in (mm)	8–20 (203–508)	3–8 (76–203)	3–8 (76–203)
Driver speed, RPM	250–600	450–1200	600–1800
Piston speed, fpm (m/s) range	750–1000 (3.8–5.1)	700–1100 (3.6–5.6)	800–1200 (4.1–6.1)
Piston speed, fpm (m/s) concentration	800–950 (4.1–4.8)	700–900 (3.6–4.6)	1100–1200 (5.6–6.1)

TABLE 12.2 Comparison of Low-, Moderate-, and High-Speed Reciprocating Compressors

as electricity—have to be brought to the package to enable it to operate. The engine burns field natural gas diverted from the compressed gas with no or minimal pretreatment. Being packaged makes the unit portable, which is quite often attractive to the upstream end user. Modern compressor packages are very sturdy and designed for installation with or without a foundation, depending on size. The skids are typically filled with concrete under all vessels, as well as in the entire engine and compressor-mounting structure. Additionally, the majority of larger skids—those in excess of ~1000 hp (746 kW)—are installed on an engineered foundation with an epoxy grout interface.

Downstream Process Reciprocating Compressors: Major Components

Discussing the typical downstream process reciprocating compressor has merit. Downstream machines are driven by an electric motor and are mounted directly onto a concrete block foundation. Electric motors are used almost exclusively because they are essentially maintenance free and have high availability. Reliability is then largely influenced by compressor-related issues, which includes the quality of associated controls ("ancillaries") and instrumentation, auxiliary support systems, and equipment installation.

We keep in mind a second option available for today's process market—the short-stroke, moderate-speed reciprocating compressor as described in Table 12.2. This is a high-speed, short-stroke compressor that is driven by an electric motor at a speed that is typically 50 to 65 percent of its full rated, upstream, natural gas engine–driven speed. This reduced speed allows the compressor to meet the reliability requirements demanded by the process end user (Refs. 4 and 5).

Crankcase. The crankcase is a U-shaped cast iron or fabricated steel frame with the top left open for installation of the crankshaft shown in Fig. 12.7. To prevent the top from flexing (opening and closing) with the forces of the throws, it is held together by torqued bolts and spacers, or alternately keyed spacers. These major stabilizers are placed directly above the main bearings.

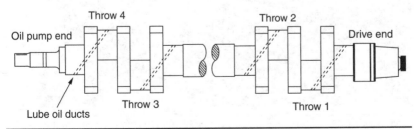

FIGURE 12.7 Typical process compressor crankshaft.

The main bearings, spaced between each throw, have removable top covers for access to the babbitted bearing shells. The keyed spacers are preferred and make access to bearing covers more easy. Main bearings for process gas compressors are often overdesigned and seldom require removal for repair. Of course, an overdesign would be of little help if dirty or inadequately formulated lubricants are used in these machines. Always use superior lubricants. They will be worth the incremental cost compared to an "acceptable" oil.

Crankshaft. The crankshaft is the central, and usually the most expensive, component of the machine. Each throw is forged and counterweights are bolted on to balance the reciprocating mass of the crosshead and piston. If the crankcase moves on its foundation, it will cause the throw to open and close through each revolution; it will then fatigue and break. For this reason, the dimension at the open end of the throw must be taken periodically while moving the crankshaft by barring it through 360 degrees. This is called *taking crankshaft deflections* and is part of a conscientious monitoring program, as outlined in Chapter 9.

Connecting rod. Connecting rods of the type shown in Fig. 12.8 are called marine type; they incorporate pretorqued bolts fastening the cap to the body at the crank end. The split is fitted with a stack of shims and one or more of these shims can be removed to compensate for bearing wear. The wrist pin is free-floating, held in place with a cap in the crosshead, which allows the connecting rod to find its own center. Figure 12.9 shows a connecting rod and two articulating power piston rods of an integral gas engine compressor.

Crosshead. The crosshead, Fig. 12.10, traverses between two crosshead guides with a clearance of ~0.001 in/in of diameter.

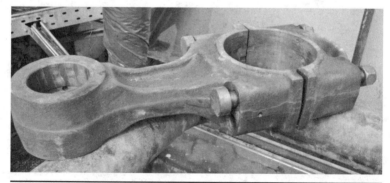

Figure 12.8 Connecting rod—moderate speed compressor. (*Source:* Field photo CES—Superior.)

FIGURE 12.9 Connecting rod and articulating power piston rods of an integral gas engine compressor. (*Source: Open Grid Europe formerly EON formerly Ruhrgas AG*, Germany, NG pipeline compressor station display.)

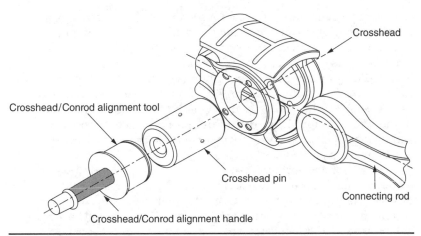

FIGURE 12.10 View of crosshead, pin, connecting rod, and alignment tool assembly. (*Source:* CES—Superior.)

Crossheads are often weighted so that the mass inertia of all reciprocating parts is sufficient to reverse the stress on the wrist pin, even when one end of the piston is under pressure. Unless load reversal (meaning stress reversal) occurs, the wrist pin will wipe the oil from the side under load. The pin will then make metal-to-metal contact and will bind. An expensive repair event will be the result.

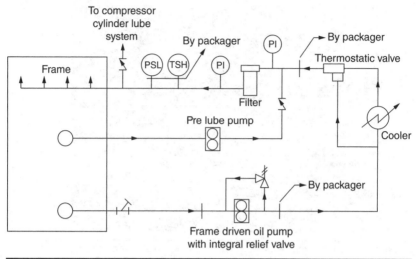

FIGURE 12.11   Reciprocating compressor lubrication schematic.

Lubrication

Lubrication of the frame is accomplished by a pump driven from the crank end, which takes oil from the crankcase sump and pumps through a cooler and a 25 micron filter, then through piping to the main bearings in the frame (see Fig. 12.11). The crankshaft has holes drilled from the main bearing surface through the connecting rod-bearing face (refer back to Fig. 12.7). From the connecting rod-bearing face, the oil passes through a hole in the connecting rod to the wrist pin and from there through holes to the crosshead sliding faces. Oil scraper rings in the frame end prevent oil leakage out along the piston rod.

Because of the tortuous passage of the oil, prelubrication is required before startup. Most specifications call for a separate motor-driven lube pump for this service. The prelube pump does not require the same flow rate as the main pump. Crankcase oil heaters are specified for outdoor compressors in cold climates to keep the oil at required viscosity and to prevent condensation which could cause corrosion. Oil, however, is a poor conductor and local overheating and carbonization has occurred when heaters with high watt-density (meaning small surface areas and high power input) were used. Therefore, when using crankcase heaters while a compressor is not in operation, the auxiliary lube pump should be continuously run.

It is important to note that the standard schematic of Fig. 12.11 shows a branch line to the compressor cylinder lube system. This would imply that the same lubricant will be used for both compressor frame and cylinders. However, decades of compressor background will not allow the coauthors to endorse this universal one-oil-fits-all approach.

Experience at best-of-class plants points to numerous instances where a different lubricant will be of great benefit to the plant's safety, reliability, and bottom-line cost. Certain synthetic lubricants have proven to be ideally suited for cylinder lubrication and these will be much preferred by reliability-focused compressor owner-operators.

Again, superior lubricants are easily cost-justified for cylinder lubrication. The extent to which these superior lubricants should also be used as the frame oil must be determined on a case-by-case basis.

Cylinders

The nomenclature and layout of a typical compressor cylinder is shown in Fig. 12.12. Here are some specifics that pertain to cylinders.

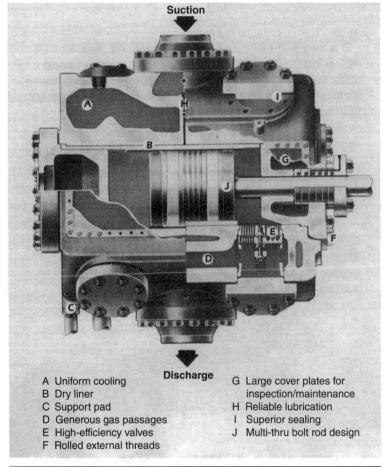

A Uniform cooling
B Dry liner
C Support pad
D Generous gas passages
E High-efficiency valves
F Rolled external threads

G Large cover plates for inspection/maintenance
H Reliable lubrication
I Superior sealing
J Multi-thru bolt rod design

Figure 12.12 Reciprocating compressor cylinder designed for long-term reliable operation. (*Source:* Dresser-Rand Company, Painted Post, New York.)

Materials of Construction

Up to a working pressure of 1000 psi, cylinders are normally made of (gray) cast iron. Above this pressure, materials of construction are cast steel or forged steel, at the manufacturer's discretion. In terms of strength, nodular iron castings fit somewhere between gray cast iron and steel, but are difficult to cast. Some owner-purchasers will not allow nodular cast iron. There is concern with potential delivery delays in the event casting difficulties are encountered by the foundry.

API-618 specifies that all cylinders have replaceable liners. These are usually made of gray cast iron because of its lubricating and bearing qualities. It is recommended that liners be honed to a finish of 10 to 20 micro in.

Cylinder Sizing

A bit contrary to our original promise not to get into mathematics, we need to consider a few calculations which will provide general background on topics such as cooling jackets, lubricant suitability, and so forth. Packing and ring material temperature limitations affect permissible cylinder and discharge gas temperatures. Also, science tells us that exceeding a certain temperature limit in the presence of oxygen and a flammable material creates an explosion hazard. This is why 375°F (190°C) is considered the maximum allowable temperature in oil-lubricated air compressors.

In somewhat general (but imprecise "isentropic") processes of compression, the discharge temperature (T_{2ad}) of a compressor may be calculated by using Eq. 12.2:

$$\frac{T_{2ad}}{T_1} = \left(\frac{P_2}{P_1}\right)^{\frac{k-1}{k}} \qquad (12.2)$$

where T_1 = suction temperature, °R
   °F = °R − 460
   P_2 = discharge pressure, psia
   P_1 = suction pressure, psia
   k = ratio of gas specific heats (c_p/c_v). It may be replaced by the "isentropic" exponent of a gas or gas mixture.

$$T_{2ad} = T_1 \times \left(\frac{P_2}{P_1}\right)^{\frac{k-1}{k}} \ [°R] \qquad (12.3)$$

A temperature of 275°F (135°C) is set as an ideal discharge temperature; a maximum of 375°F (190°C) is considered the absolute limit of operating temperatures. The design ratio per cylinder (or stage) can be determined by using Eq. 12.3 while plugging in the above limits.

An additional limit is presented by the design compression and tension load on the piston rod. Limiting rod loads are set by the compressor designer or the original equipment manufacturer (OEM). These loads must not be exceeded and, therefore, rod load must be checked on each cylinder application. Rod load is defined as follows in Eq. 12.4:

$$RL = P_2 \times AHE - P_1 \times ACE \qquad (12.4)$$

where RL = rod load in compression, lb
AHE = cylinder area at head end in any one cylinder, in^2
ACE = cylinder area at crank end, usually ACE = AHE − area rod, in^2
P_2 = discharge pressure, in psia, and P_1 = suction pressure, psia

Some manufacturers require lower values on the rod in tension than compression. If this is the case, the above limits must be checked by reversing the expressions AHE and ACE. However, for most applications, the compression check will govern.

Compressors below 500 psig, such as plant air compressors, are usually sized by considering discharge temperatures. Rod load often becomes the limiting factor on applications above these pressures.

Cylinder Cooling

Excessive cooling, also called "overcooling," of cylinders can cause condensation, which then leads to lubrication deficiencies and corrosion. For this reason a thermosiphon system as shown in Fig. 12.13 is preferred. This means that no forced coolant circulation ought to be applied wherever possible. However, high suction temperatures or high compression ratios do require coolant circulation.

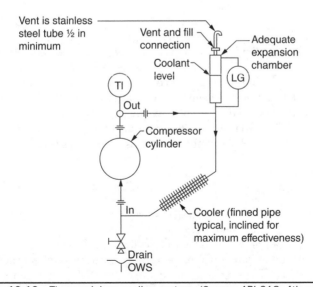

Figure 12.13 Thermosiphon cooling system. (*Source:* API 618, 4th ed.)

Thermosiphon cylinder cooling may be specified with a discharge temperature limit of about 200°F or 93°C. Thermosiphon cooling consists of filling the jackets with an appropriate liquid such as water or a light heat transfer oil and letting the heat radiate from the outer cylinder walls. The purpose of filling the jackets is to obtain even heat distribution throughout the cylinder.

Above 200°F or 93°C, coolant circulation is applied. Raw water should not be used because it leaves deposits in the cooling jackets, which are extremely difficult to clean out. A closed system is therefore frequently specified. This system consists of a reservoir with suitably treated water, circulating pumps, and one or more heat exchangers.

Overcooling of the cylinder to a temperature below the dew point of the compressed gas must be avoided; overcooling will cause cylinder corrosion, lubricant wash-off, or liquid build-up. The presence of liquids of any kind is likely to cause "slugging," the intermittent and unpredictable ingestion of a slug of liquid. Therefore, temperature controllers must be installed and excess coolant bypassed around the cooler or heat exchanger. It is also recommended that a thermostat and heater be mounted on the reservoir and the coolant circulated when the compressor is stopped or on standby. The purpose, of course, is to maintain the cylinders at a temperature above gas dew point.

The coolant is usually 50 percent propylene glycol and water. This mixture will prevent freezing, but has a lower specific heat than water. Therefore, the manufacturer must be asked to size heat exchangers on the basis of lower specific heat, even if the initial coolant is intended to be only water.

Valves

Typical compressor valves are shown in Figs. 12.14 through 12.16. Valves are the most critical parts of a reciprocating compressor. They are sensitive to both liquids and solids in the gas stream, causing plate and spring breakages. When the valve lifts, it can strike the guard and rebound to the seat several times in one stroke. This effect is called valve flutter and it leads to breakage of valve plates. The problem is predominantly found with light molecular weight gases such as hydrogen. The problem can be mitigated by restricting the lift of the valve plate, thus controlling valve velocity. Valve velocity is specified as (Eq. 12.5):

$$V = \frac{D \times 144}{A} \quad \text{ft/min} \tag{12.5}$$

where V = average velocity in ft/min
D = cylinder displacement in cubic ft/min
A = total inlet valve area per cylinder, calculated by valve lift times valve opening periphery, times the number of suction valves per cylinder in square inches. Valve lift distances differ with molecular weights and are shown in Table 12.3.

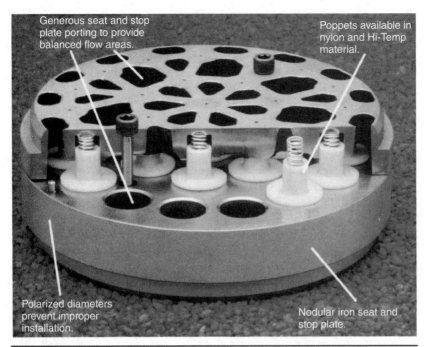

FIGURE 12.14   Poppet valve. (*Source:* Cooper Cameron Corporation, Cooper-Bessemer Reciprocating Products Division.)

FIGURE 12.15   High-flow ring-type discharge valve. (*Source:* France Compressor Products.)

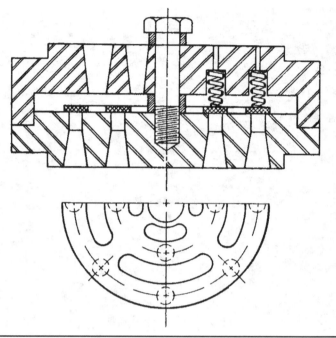

FIGURE 12.16 Slotted disc valve. (*Source:* CES.)

Gas Mol. Weight	Valve Lift in (mm)
<7	0.040 (1.00)
7–20	0.060 (1.52)
>20	0.080 (2.00)+

TABLE 12.3 Typical Values for Compressor Valve Lifts

Later, in Chapter 13 (Fig. 13.8, dealing with valve loss factors) it will become obvious that lower valve lifts and consequently higher valve velocities result in increased valve power losses. However, lower valve lifts result in a more reliable valve and designers, owners and operators of reciprocating compressors must therefore, find the balance between valve reliability and valve losses.

The typical upstream high speed, high piston speed, short-stroke compressor uses high-lift, high-efficiency valves to minimize the reduction in efficiency resulting from the high piston speed. The valve plate material, lift, and springs are specifically selected to fit the application.

While poppet and ring valves are primarily applied in moderate-to-slow speed compressors, plate valves are predominantly used in high-speed machines. Plate valves, at present, appear to be the most promising candidates for further development as the trend toward higher compressor speeds continues.

Manufacturers sometimes have interchangeable suction and discharge valves. This can lead to putting valves in the wrong port, which can result in massive valve breakages or broken rod or cylinder. It is therefore necessary to specify that valves must not be interchangeable. However, certain non-interchange features can be lost or broken off, which means that correct valve placement should always be checked.

Pistons

Pistons are usually cast iron and are often of hollow construction so as to reduce weight. The void space can fill with gas and is an explosive hazard when removing the piston from the rod. One must, therefore, specify that an easily removable plug be supplied to vent this space before handling. Larger pistons are made of aluminum to reduce weight. These pistons have large clearance in the bore to allow for thermal expansion, in the order of 20 mils (0.020″)/in of diameter. Wear bands or rider rings are often supplied on cast iron pistons and are necessary on all aluminum pistons, see Fig. 12.17. Bearing loads of pistons and rider rings are based on a unit stress

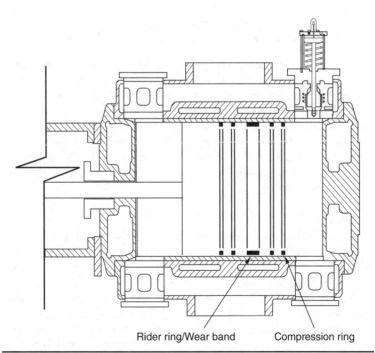

Rider ring/Wear band   Compression ring

Figure 12.17 Rider ring/wear band placement on a piston.

value and may be calculated using Eq. 12.6. Normal limits are 12 psi (0.08 N/mm²) for cast iron, 14 psi (0.1 N/mm²) for bronze, and 22 psi (0.15 N/mm²) for one of the very many low-oxidation alloys, known as Allen metal.

As long as the maximum average piston speeds remained in the 800 to 850 fpm (4.1–4.3 m/s) range, the use of wear bands by high-speed, short-stroke compressor manufacturers was not common. Pushing the maximum piston speed into the 1100 to 1200 fpm (5.6–6.1 m/s) range made the use of wear bands a firm requirement to avoid rapid wear of the compressor cylinder or liner main bore. Wear band–bearing pressure loads in the range from 10 to 12 psi (0.07–0.08 N/mm²) actually provide acceptable wear band life in the typical upstream application. However, many end users and manufacturers have found this wear band life inadequate for downstream applications where higher reliability is desired. Here, wear band–bearing pressure loads in the 5 to 6 psi (0.035–0.040 N/mm²) range are used.

$$\text{Wear band–bearing pressure} = \frac{\text{weight}}{C \times D \times W} \quad \text{psi (N/mm}^2) \quad (12.6)$$

where weight = weight of the complete piston assembly plus ½ the weight of the piston rod, lb_m (kg)
C = 0.866 (0.0883)
D = piston diameter, in (mm)
W = total width of all wear bands on the piston, in (mm)

Teflon or Teflon component compression rings are often specified and are useful up to 500 psi pressure differential (ΔP). Above these loadings, copper-bearing alloy material or babbitt are used for wear rings and plain bronze for compression rings. Teflon compression rings are often offered with steel expander rings underneath. Still, these should be avoided because, as the compression rings wear, the expander rings can score the cylinder. Pistons and rings designs are available that will hold the compression ring out against the cylinder without expander rings.

Piston Rod

The piston rod screws into the crosshead; it must be locked to prevent backing off, either by a locknut or a pin. However, there are several other methods of fastening the rod, and they should be investigated as part of a prepurchase reliability assessment (see our chapter on MQA) of a reciprocating compressor. The rod is adjusted in the crosshead to equalize the end clearance of the piston in the cylinder. End clearance is checked by barring over the machine and crushing a strip of soft lead. The remaining thickness is then measured; it is commonly referred to as the bump clearance.

The rod must be hardened where it passes through the packing. Some rods are chrome plated, but problems have occurred, especially on high-pressure machines with severe heat buildup, causing spider web cracks in the chrome which in turn can flake off and destroy the packing. Colmonoy plating is advantageous, but often the best arrangement is to purchase flame-hardened rods. Hardened rods, when worn, can be plated with tungsten carbide, which tends to last the life of the machine. High-pressure machines often have the rod extended through the piston and out of the cylinder head to balance the pressure load on the piston, i.e., the rod load. This arrangement is referred to as a tail rod. Tail rods have been known to break off and get ejected from the cylinder like a missile. Reliability-focused owner-purchasers specify that all tail rods be housed in a containment guard strong enough to contain the tail rod should a breakage occur.

Packing

Segmental ring packing is a standard component used to seal the space between the compressor cylinder and the piston rod, see Fig. 12.18. It consists of circular cups, usually cast iron, with their open ends away from the cylinder. Each cup contains a ring cut tangentially from ID to OD at outside of cup and another ring cut radially from ID to OD

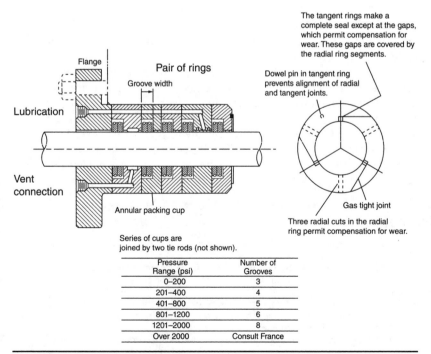

Pressure Range (psi)	Number of Grooves
0–200	3
201–400	4
401–800	5
801–1200	6
1201–2000	8
Over 2000	Consult France

Figure 12.18 Packing assembly. (*Source:* France Compressor Products.)

inside the cup. Each segmented ring is held together with a garter spring. The flat side of each cup and the flat sides of the rings are lapped to a very fine finish. The cups are bolted together in a stack; the number of cups depends on the pressure to be sealed. Above 500 psi, an additional metallic back-up ring is often used to prevent Teflon—a favorite elastomer—extruding from the cups.

Rings can be manufactured from many materials, but the most common are Teflon, tinned iron, and bronze. Again, our preference is for Teflon-based materials whenever practicable.

The stack of packing rings has a vent; oil supply and drain holes are drilled in the packing case at appropriate locations. The rings must be correctly aligned each time the packing case is opened.

Compressor manufacturers supply a distance piece between the cylinder and the crankcase for access to the packing. This is usually sufficient for a three or four cup packing set. In process applications, however, especially above 2000 psi or 140 bar, packing sets can amount to 8 or even 18 cups. Extra long-distance pieces will be needed in order that sufficient space is available to open the cups and change the packing rings. If any doubt exists, the extra long-distance piece should be specified at original purchase; the additional or incremental cost is minor. It is almost impossible to change or upgrade to larger packing sets after the machine is built.

What We Have Learned

Small single-acting reciprocating compressors differ from process reciprocating machines.

Different machine categories operate at different piston speeds. Machines with higher piston speeds require more maintenance and have lower availability than slower speed machines.

Compressor frame lubrication and cylinder lubrication are different sources or circuits. Each usually merits a different lube type in reliability-focused plants.

Developing a sound working relationship with competent, dependable suppliers and service contractors is important.

References

1. Ingersoll-Rand and A.W. Loomis, eds, Compressed Air and Gas Data, 3rd ed, 1980.
2. American Petroleum Institute, Alexandria, VA; API Standard 618, *Reciprocating Compressors for Petroleum, Chemical, and Gas Industry Services*, 5th ed, December 2007.
3. American Petroleum Institute, Alexandria, VA; API Specification 11P *Specification for Packaged Reciprocating Compressors for Oil & Gas Production Services.*
4. *CompressorTech2*, Data Handbook 2010.
5. Ariel Corporation, Mount Vernon, Ohio, www.arielcorp.com.

CHAPTER 13

Compressor Operation and Capacity Control

Starting Reciprocating Compressors

The work of compression is actually done in less than 180° of each rotation. Therefore, rotating inertia must be built into the power train to obtain a uniform power draw from the prime mover. The stored energy must be put into the system before compression starts; otherwise the prime mover will be overloaded.

The compressor is unloaded to start, either by suction valve unloading or through a valve bypass from discharge back to suction.

A popular means of suction valve unloading consists of fingers which hold open the suction valve plates. Another unloader style consists of a plug covering a hole in the center of the suction valve. Either may be actuated by a pneumatic or electrical plunger, or by a hand wheel mounted on a screw spindle.

Finger-style suction valve lifters are normally used in air compressors. Suction valve plugs are much preferred for process compressors. Finger-type suction valve lifters put point loads on valve plates when actuated, causing a stress concentration. Plug valves offer a port area which may be less than the valve area, thus causing some throttling. Still, they require fewer maintenance interventions than finger-type valve lifters in corrosive services.

A valved bypass is normally provided for starting single-stage compressors. This arrangement would not be suitable for multistage machines since their first stage would load up due to throttling in the smaller stages that follow. Actuators are normal for multistage compressors but they must be timed to operate in unison such that one stage will not carry the entire load at start-up.

Performance Control and Throughput Adjustment

Performance and throughput control can be achieved by unloaders. Actually, compressor unloaders are commonly used for two primary purposes:

- Unloading the compressor cylinders during compressor start-up. This helps minimize the necessary starting torque and thus reduces the size and cost of the compressor driver.

- Allowing reciprocating compressor to meet a variety of different gas flow conditions through the use of unloaded compressor cylinder ends and/or clearance pockets.

In order to increase cylinder end clearance and thus decrease capacity, valved pockets (Fig. 13.1) can be added to the cylinder volume. They are referred to as variable volume clearance pocket (VVCP) unloaders and are used to open and close a clearance volume (a "pocket") inside the compressor cylinder. Opening a clearance pocket adds clearance volume to the compressor cylinder, decreasing compression efficiency and resulting in decreased gas flow. Use of clearance pockets can help obtain intermediate capacity load steps not achievable by the sole use of cylinder end unloading.

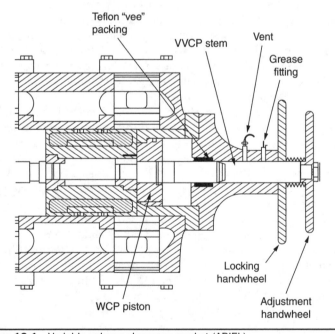

FIGURE 13.1   Variable volume clearance pocket (ARIEL).

Variable volume clearance pocket unloaders present the main means of capacity control on upstream high-speed compressors. In upstream applications, VVCPs are placed in the cylinder end cover, or on the side of a cylinder, and the valves to open them can be operated by a hand wheel or actuator. On compression ratios below 2, the volume of the pockets becomes too large to be effective, and only partial reduction of cylinder end capacity control is accomplished by VVCPs. On ratios below 1.5, the volume of an effective pocket becomes prohibitive and pocket control is rarely used. Variable volume clearance pockets can thus give infinite modulation over the range of the pocket. However, the method is considered cumbersome for downstream process operations, and is therefore, not very often used except in special cases where fine control is mandatory.

Several other styles of unloaders are primarily used on process gas compressors. They include:

The port unloader. A port unloader is an unloader located in the compressor cylinder such that it can be used to open and close a dedicated "port." This port connects the cylinder bore with the internal cylinder gas passage. Opening the port will allow gas to flow freely into and out of the cylinder. This will then effectively unload the cylinder end in which the port unloader is located. Port unloaders are commonly used in cylinders where there is sufficient space for an unloader port in addition to the required number of compressor valves.

A plug unloader. Shown in Fig. 13.2, a plug unloader is used in conjunction with a partial inlet compressor valve. When the unloader is "loaded," it seals off the open hole in the center of the compressor

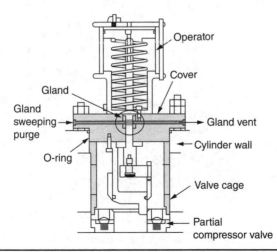

FIGURE 13.2 Valve plug unloader. (*Source:* Dresser-Rand.)

valve and the valve then operates normally. When the hole in the center of the valve is opened or "unloaded," gas can pass freely into and out of the cylinder, unloading the cylinder end in which the unloader is located. Plug unloaders are commonly used where there is insufficient space for a port unloader, but where proper unloading can be achieved through partial compressor valves.

A finger unloader. A finger unloader uses "fingers," as shown in Fig. 13.3, to depress the inlet compressor valve elements and to

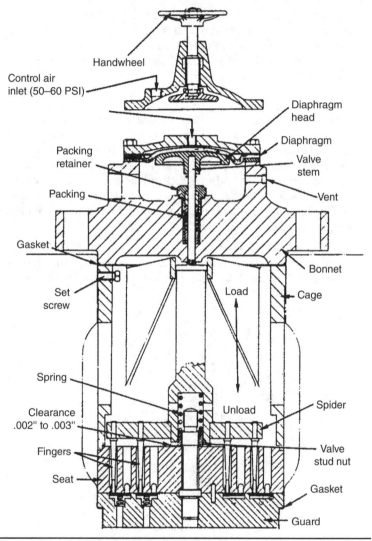

Figure 13.3 Suction valve (finger) unloader. (*Source:* Cooper Industries.)

hold them open. Doing so allows the gas in the compressor cylinder to pass freely into and out of the compressor cylinder. Finger unloaders are commonly applied in situations where, because of insufficient unloading area, port and plug type unloaders cannot be applied.

A suction valve lifter. Suction valve lifters or plugs are often used in air process gas services. They unload a specific end of a cylinder completely. Control is, therefore, in steps. On a single double-acting cylinder, this is called three-step control, giving capacities of 0, 50, and 100 percent of rating. Suction valve unloading can be combined with clearance pockets to obtain five-step unloading. Five-step unloading is accomplished by having a clearance pocket in one end of the cylinder to reduce capacity by 25 percent. This would result in a capacity arrangement of capacities of 0, 25, 50, 75, or 100 percent of rating.

Infinite modulation can also be obtained by a device which times the closing of the suction valve partway down the compression stroke. As of 2012, this is a promising but somewhat expensive electronic control offered by at least one major compressor valve manufacturing company.

Suction pressure control is often recommended where temperature rise and rod load limits allow it. The performance of the specific machine must be checked before applying this method. An important detail to note is that a compressor operating at constant discharge pressure with variable suction pressure will reach its peak horsepower approximately at a compression ratio of 2.2. Therefore, at or near this ratio, variable capacity at relatively constant horsepower can be obtained. On some applications, where variable suction pressures are built into the application, clearance pockets are applied to keep the peak load below the driver horsepower.

Multiple Services

The "running gear," composed of reciprocating compressor frame, crankshaft, gear (occasionally used for speed reduction), and driver represents the greater part of the cost of these machines. The cost of the cylinder and piston is relatively small. For this reason, the combination of more than one service on a given frame can result in substantial savings in the total cost of compression facilities. A typical example would be the placement of feed and recycle gas cylinders on one common frame for a hydrofiner unit in a petroleum refinery. In this instance, the saving in cost for compressor and driver would be about 40 percent compared to the cost of two separate machines.

Calculations

The purpose of showing calculations in our text is to help understand and troubleshoot reciprocating compressor problems. The math just identifies how some relationships are derived. Usually, we want to:

1. Determine the approximate power requirement to compress and move a certain volume of gas from some intake conditions at a given discharge pressure.

2. Ascertain the capacity of an existing compressor installation under field suction and discharge pressure conditions.

Solving compressor problems using the ideal gas laws can be reduced to a relatively simple sequence of applying a few fundamental formulas and obtaining values from simple curves. In order to help in the understanding of these terms and equations, the first portion of this section covers some basic thermodynamics.

Following the sequence of events, as they happen in a cylinder, is probably the best way to develop an understanding of the terms that are likely encountered in most compressor problems. We are helped by referring to Fig. 13.4, where the typical pressure–volume relationship is superimposed on a piston with a stroke length of "s."

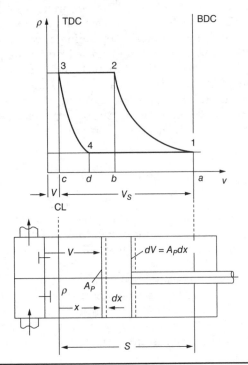

FIGURE 13.4   Ideal compression cycle (P–V diagram).

Position 1: This is the start of the compression stroke. The cylinder has a full charge of gas at suction pressure. As the piston moves toward Position 2, the gas is compressed along line 1-2.

Position 2: At this point, the pressure in the cylinder has become greater than the pressure in the discharge line. This now causes the discharge valve to open and allows the original charge of gas, which is now at discharge pressure, to be moved into the discharge line.

Position 3: Here, the piston has completed its discharge stroke. As soon as it starts its return stroke, the pressure in the cylinder drops, which closes the discharge valve. Notice the volume of gas that is trapped between the end of the piston and the end of the cylinder. This is known as the clearance volume. This volume of gas has to expand along line 3-4 as the piston moves back toward Position 4.

Position 4: At this point, the pressure within the cylinder has dropped below the suction pressure, causing the suction valve to open. This allows a new charge of gas to be taken into the cylinder as the piston returns to Point 1, the start of the compression stroke.

With the basic mechanics of the compressor in mind, it becomes easy to visualize and understand most of the basic definitions that will be offered here.

Piston Displacement

Piston displacement (PD) is the actual volume displaced by the piston as it travels the length of its stroke, from Position 1, bottom dead center (BDC), to Position 3, top dead center (TDC). Piston displacement is normally expressed as the volume displaced per minute or cubic feet per minute (CFM). In the case of the double-acting cylinder, the displacement of the crank end of the cylinder is also included. The crank end displacement is, of course, less than the head end displacement by the amount that the piston rod displaced. For a single acting cylinder, Eq. 13.1 expresses the piston displacement PD:

$$PD = \frac{A_{HE} \times S \times RPM}{1728} \quad [CFM] \qquad (13.1)$$

where A_{HE} = area head end of piston, in^2
S = stroke, in

For double-acting cylinders, the piston displacement PD is expressed in Eq. 13.2:

$$PD_{DA} = \frac{A_{HE} \times S \times RPM}{1728} + \frac{A_{CE} \times S \times RPM}{1728} \quad [CFM] \qquad (13.2)$$

or, also acceptable, per Eq. 13.3:

$$PD_{DA} = \frac{2S \times RPM}{1728} \times \left(A_{HE} - \frac{1}{2} A_R \right) \quad [CFM] \qquad (13.3)$$

where A_R = rod area, in^2

Compression Ratio

P_2/P_1 is the ratio of the discharge pressure to the suction pressure with both pressure values expressed as absolute units. In the P–V diagram of Fig. 13.4, line 2-3 is at discharge pressure, and line 4-1 is at suction pressure.

Note that the compression ratio is not controlled by the compressor, but by the system conditions on the compression suction and discharge. During compression, the cylinder pressure will rise until it is high enough to force the gas through the discharge valves and into the system. During expansion, the cylinder pressure will fall until the system can force gas into the cylinder and maintain the pressure. Naturally, the compression ratios cannot be infinite. As the compression ratio is increased, volumetric efficiency will decrease until no more gas is moved. If the compression ratio is increased at constant suction pressure, the piston rod loadings will increase and the overloaded piston rod may fail. Further, the cylinder cooling provided may not be adequate for the discharge temperature associated with higher compression ratios. Nevertheless, the machine will always attempt to meet the system compression ratios.

On multistage compressors, the actual compression ratios on each stage are self-determining because all the gas from the low pressure cylinders must also pass through the higher pressure cylinders. The compression ratio on each stage will not change in proportion to the system pressure ratio, but it can change appreciably if the efficiency of any of the stages undergoes change. Efficiency will decay if there is piston ring blow-by or valve failure. Perhaps the most elusive cause of efficiency decay is erosion and blow-by at the valve seat. Although the seating surface is thought to be sealed by a gasket, such erosion is not usually suspected by maintenance personnel.

Clearance

Percent clearance is used to calculate the volumetric efficiency of a cylinder and is the ratio of clearance volume to piston displacement, expressed as a percentage, per Eq. 13.4:

$$\% \ CL = \frac{\text{clearance volume } (in^3)}{\text{piston displacement } (in^3)} \times 100 \qquad (13.4)$$

Clearance volume is the volume left in the cylinder end at Position 3 of the stroke shown in Fig. 13.4, including the valves and valve ports.

Clearance volume is the region in each end of a cylinder that is not swept by the piston movement. Generally, large cylinders are affected more by the clearance in the gap between piston and cylinder head. In contrast, small cylinders are affected more by clearance in and under the valves. Clearance percentage varies between each end of the cylinder and should be considered separately in calculations. It can usually be averaged. Percent clearance is vital information for reciprocating compressor calculations and therefore should be specified as required nameplate information on all compressor purchases. For estimating purposes and if clearance is unknown, assuming a 15 percent clearance will yield reasonable accuracy.

Volumetric Efficiency

Volumetric efficiency (EV_S) referred to suction conditions is the ratio of actual gas drawn into the cylinder to the piston displacement, expressed in percent. Referring again to the diagram in Fig. 13.4, volumetric efficiency represents the distance 4 to 1, divided by the stroke. Theoretically, this is expressed by Eq. 13.5 as:

$$EV_S = 1 - CL\left[\left(\frac{P_2}{P_1}\right)^{\frac{1}{k}} - 1\right] \tag{13.5}$$

where in actual machines, other factors affect the volumetric efficiency. These are leakage across the valves, across the piston rings, and through the packing. There is also the heating effect on the incoming gas from the residual heat in the cylinder. Because, in addition, the compressibility of the gas comes into play, a better formula has been developed and is shown in Eq. 13.6.

$$EV_S = 1 - L - CL\left[\frac{Z_1}{Z_2}\left(\frac{P_2}{P_1}\right)^{\frac{1}{k}} - 1\right] \tag{13.6}$$

where Z_1 = compressibility factor at compressor inlet conditions, signifying the deviation of the gas or gas mixture from the ideal gas laws.
Z_2 = compressibility factor at compressor discharge conditions, signifying the deviation of the gas or gas mixture from the ideal gas laws.
L = Loss correction, as a decimal, is taken from Fig. 13.5.

Except for extreme cases, compressibility factor has a minor effect and is usually ignored. Curves have been developed for Eq. 13.7, from which "L" may be subtracted and give satisfactory results for most applications. However, it should be noted that several versions

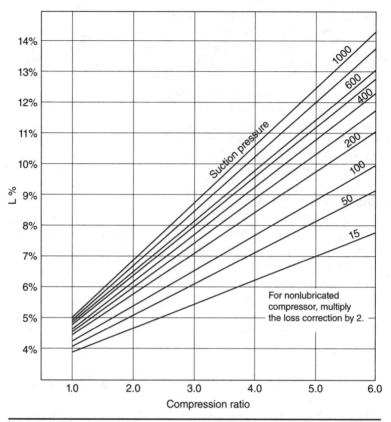

Figure 13.5 "L" as a function of compression ratio.

of volumetric efficiency equations exist. While Eq. 13.6, above, is a theoretical approach to determining volumetric efficiency, reciprocating compressor design and original equipment manufacturing (OEM) companies have developed their own formulas. One well-known practical version is expressed in Eq. 13.7:

$$ EV_S = 0.97 - 0.008 \times \left(\frac{P_2}{P_1}\right) \times \left(\frac{P_1}{P_b}\right)^{0.2} - CL \times \left[\left(\frac{P_2}{P_1}\right)^{\frac{1}{k}} - 1\right] \qquad (13.7)^a $$

where P_b = base pressure in psia, normally 14.7

Figure 13.6 illustrates the effect of clearance on compression ratio and volumetric efficiency.

---

[a]Cooper Bessemer Formula

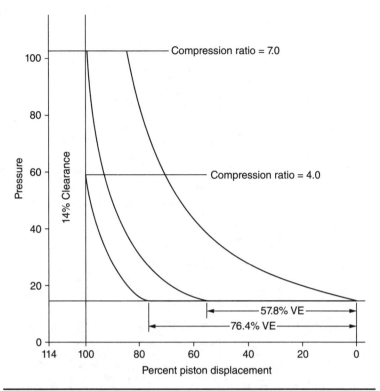

FIGURE 13.6   Volumetric efficiency as a function of compression ratio and clearance at moderate and high compression ratio conditions. A P-V diagram for a ratio of 7 is superimposed on a diagram for a ratio of 4, all else being the same. (*Source:* Dresser-Rand Company, Painted Post, N.Y.)

Basic Equations

Reciprocating compressor calculations are performed by stage. Therefore, the number of stages must be determined first. The number is based on limiting temperature and rod load, as was discussed earlier. In addition, horsepower savings can be realized by using the same ratio for each stage. This ratio can be calculated by Eq. 13.8:

$$\text{Ratio per stage} = (\text{overall ratio})^{\frac{1}{\text{number of stages}}} \qquad (13.8)$$

Generally speaking, in process work, the ratio per stage seldom exceeds 3. Equation 13.9 is the basic equation for reciprocating compressor horsepower:

$$\text{HP}_{\text{AD}} = \frac{\text{ACFM} \times 144 \times P_1}{33000} \times \frac{k}{k-1} \times \left[\left(\frac{P_2}{P_1} \right)^{\frac{k-1}{k}} - 1 \right] \qquad (13.9)$$

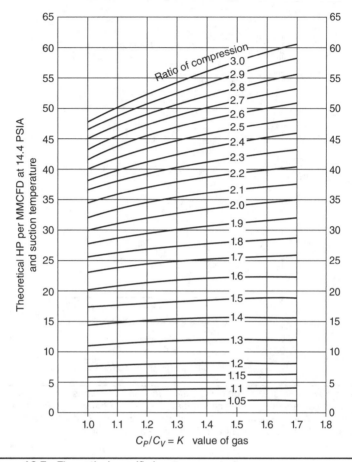

FIGURE 13.7   Theoretical specific horsepower.

This equation has been modified and presented as a curve, making it easier to visualize the various relationships and influences. Million cubic feet per day at 14.6 psia and suction temperature are displayed and presented in Fig. 13.7. The horsepower is theoretical and must be modified by taking compressibility, valve efficiency, and mechanical efficiency into account.

Valve efficiency allows for pressure drop across the valves, which results in higher discharge and lower suction pressures within the cylinder, than is supplied to the compressor. This may be found from Figs. 13.8 and 13.9; again, it shows how changing a parameter influences the results.

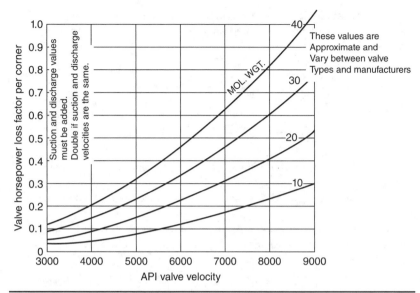

FIGURE 13.8   Valve power loss versus valve velocities.

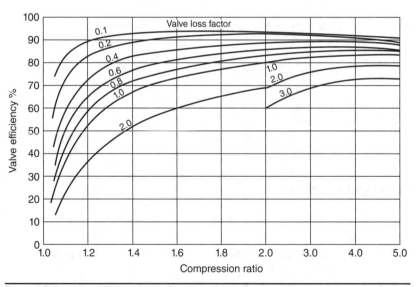

FIGURE 13.9   Valve efficiency as a function of compression ratio and valve loss factor.

Mechanical efficiency is based on the power loss in the crankcase, and efficiency is usually accepted as 95%. Then, actual horsepower may be calculated using Eq. 13.10:

$$HP_{Actual} = \frac{MMSCFD \times T_1}{Z_{STD} \times T_{STD}} \frac{BHP}{MMCFD} \frac{(Z_1 + Z_2)}{2}$$

$$\times \frac{1}{valve\ efficiency} \times \frac{1}{mechanical\ efficiency} \quad (13.10)$$

where MMSCFD is a million standard cubic feet per day at 14.4 psia and 60°F or ~520°R.

This equation is used for finding horsepower before sizing a machine.

For existing machines, where displacement is known and volumetric efficiency can be calculated, Eqs. 13.11 and 13.12 apply:

$$MMSCFD = 0.00144 \times \left(\frac{P_1}{P_b}\right) \times \left(\frac{T_b}{T_1}\right) \times \left(\frac{Z_b}{Z_1}\right) \times PD \times EV_S \quad (13.11)$$

$$where\ 0.00144 = \frac{1440\ minutes/day}{10^6\ scf/MMSCF}$$

or

$$HP_{Actual} = Displacement \times \frac{BHP}{MMCFD} \times \frac{EV_S \times P_1}{10^4} \times \frac{Z_1 + Z_2}{2}$$

$$\times \frac{1}{valve\ efficiency} \times \frac{1}{mechanical\ efficiency} \quad (13.12)$$

where P_b = base pressure in psia, normally 14.7
T_b = base temperature in degrees Rankine, normally 520
Z_b = base compressibility, normally 1.0

What We Have Learned

- Unloaders are used for compressor start-up and for capacity control.
- There are several unloader styles and each has its advantages and disadvantages.
- Multiple services are feasible and should be considered in process applications. They can represent major cost savings.
- Volumetric efficiency is an important property built into a compressor cylinder. It should influence compressor selection.

Bibliography

2011 Ct2 Compressor technology sourcing supplement, 2011 ed., CompressorTechTwo, Diesel and Gas Turbine Publications, 2011.

Phillippi G., and J. Spiller, Short stroke reciprocating compressors—high-speed upstream vs. moderate speed downstream, CompressorTechTwo, January-February 2011, 34–42, Diesel and Gas Turbine Publications, 2011.

Spiegel, B., M. Testori, and G. Machu, Next-generation valve technology for high-speed compressors. CompressorTechTwo, April 2011, 70–79, Diesel and Gas Turbine Publications, 2011.

Bloch, Heinz P., and J.J. Hoefner, Reciprocating Compressors: Operation and Maintenance, Gulf Publishing Company, Houston, TX, 1996.

CHAPTER 14

Reciprocating Compressor Maintenance

A few pages are needed to guide owners and operators of reciprocating compressors in setting up maintenance and trouble shooting procedures for their machines. Our recommendations here are not intended to replace the stipulations of the original equipment manufacturer (OEM) for any machine. Instead, they should serve as general guidelines to assist in failure prevention and problem solution.

Recent investigations showed maintenance costs of around US$ 48.00 per horsepower per year for reciprocating compression equipment as opposed to some US$ 10.00 per horsepower per year for large turbocompressors. Despite the application of advanced technologies in the development and manufacture of wearing parts, such as valves, piston rings, and packing, these components continue to be the main causes of reciprocating compressor failures and unscheduled outages. Figure 14.1 shows the most frequently failing reciprocating compressor components.

Compressor Cylinder

Rod Load

Major breakages such as broken rods, crossheads, pin bushings, and frame cracks are caused more often by exceeding maximum rod load than any other cause.

Massive failures become more likely on multistage machines and cylinders with higher suction pressures; these failures can occur with frightening suddenness. Specific causes are failure of interstage relief valves or shutdown switches, valve breakage, and unequal loading. These malfunctions can relate to opening clearance pockets or valve

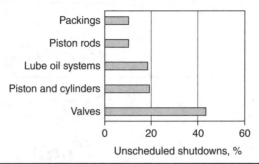

Figure 14.1   Reciprocating compressor failure causes.*

lifters out of sequence and changing process conditions such as lowering suction pressure without checking if the resulting loads remain within design range.

A monitoring curve should be drawn for each cylinder where excessive rod loads could occur. The boundaries of the curve(s) will be discharge pressure relative to suction pressure. Two straight-line curves may be obtained from the following two equations:

$$P_2 = \frac{\text{CRL} + P_1(A - a)}{A} \tag{14.1}$$

$$P_2 = \frac{\text{TRL} + P_1 A}{A - a} \tag{14.2}$$

where P_1 = suction pressure in psi
  P_2 = discharge pressure in psi
  A = cylinder area in in^2
  a = rod area in in^2
  CRL = permissible compression rod load in lb
  TRL = permissible tension rod load in lb

By substituting two reasonable values for P_1 in each equation, two corresponding points are calculated for P_2. A straight line is drawn joining these two points. The lower of the two lines will govern. When only one rod load is given, the compression rod load formula should be used.

In essence, for every discharge pressure there exists a corresponding limiting suction pressure. Any particular possible pressure should be checked on such a curve; it will look similar to Fig. 14.2. The combination of suction and discharge pressures allowed in a particular compressor must fall below the lower of the two lines.

Some operators prefer this curve be made into table form as shown below. Extracting the curve above onto a table is a simple matter. See Table 14.1.

---

*Refinery experience.

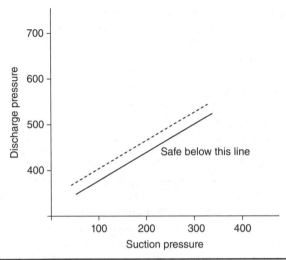

FIGURE 14.2 Rod load limit.

Keep Pressures	
Suction (psi) above	Discharge (psi) below
150	380
200	420
250	460
300	510
350	550
400	590

TABLE 14.1 Suction vs. Discharge Pressure Limits

Piston and Liner Care

Cooling Issues

Overcooling of cylinders can cause vapor condensation and can rapidly lead to loss of lubrication and corrosion. Care should be taken that the circulating coolant is above the dew point of the gas being compressed. Air compressors are the most prominent offenders because usually the cylinder cooling jackets are piped in series after the intercooler. The flow rate through the jackets should be reduced such that the coolant is warm to the touch at outlet, or about 110°F (43°C).

Temperature Issues

Cylinder discharge temperatures are the most sensitive readings that can be observed on a compressor. Any cylinder operating at constant pressures that shows a rise-over-baseline of 25°F (14°C) or more in discharge temperature is cause for concern. The machine should be shut down immediately and checked for broken valves, piston ring failures, and/or liner scoring.

Rider Rings

Pistons with rider rings or wear bands must be checked for rider ring (also called rider band) wear. There are two types of rider rings, fixed and floating. With fixed rider rings, the piston must be rotated 90° every so often to equalize wear. With floating riders, they should rotate on the piston on their own and therefore equalize wear. In either instance, measuring compressor rod run-out with a dial indicator at the cylinder gland is productive. While barring over, deviation measurements will indicate whether or not rider bands should be rotated or replaced. The clearance under the piston should be measured on installation and compressor rod run-out readings recorded. When, with time, the rod drop change adds + 0.005" (0.13 mm) to this clearance, rider or wear rings should be rotated or renewed. Rod drop (rod sag) readings can be taken with the compressor operating; this monitoring method is simpler than using feeler gages under the piston. Taken quarterly and plotted on a time line, rod drop will indicate replacement time under normal wear as well as unusual wear conditions. Information about rod drop monitoring procedures may also be found in *API Standard 618*, 4th edition, June 1995, Appendix C. Of course, more recent editions of this API standard must be considered whenever available.

Rod Drop Detection

While the above describes traditional checks of piston ring wear, permanently installed rod-drop devices have gained widespread acceptance. These are composed of a proximity probe which, through the use of an eddy current and a calibrated gap, monitors continuously any change in piston rod up-down travel. This travel is likely to occur as a result of packing or ring wear and/or rod deflection due to overload. Rod drop recording can result in increased machine availability and improved maintenance scheduling.

Pistons

Most pistons today (2012) are using Teflon or composition type compression rings. Compositions are predominantly high-performance polymers (HPPs). The manufacturer's experience with proposed compression ring materials in a particular gas service must be checked before purchase.

Frequently, compression rings are forced out against the liner by steel expander rings. If the compression rings should become worn and break up, the expander ring can expand out and damage the liner which will create a bad cylinder score in a very short period.

Scheduled inspection of this type of compression rings through a valve port is essential. This can be done on some types of compressor cylinders by backing the rod off from the crosshead, moving the piston rings under the valve port, and measuring the end gap clearance of the rings. It is a good rule-of-thumb not to install piston ring expanders or wave springs under compressor piston rings that are in excess of 10 in (10") (254 mm) in diameter.

Additionally, the authors would like to stress the need for running-in newly installed wear components such as the rings mentioned above. There is a large body of evidence acknowledging a temperature increase during the initial hours of unloaded operation. During this time, thermoplastics release their lubricant compounds by coating the mating part as mentioned before. A 4 to 6 hour run-in is appropriate.

Piston Rings
Metallic compression rings, if they must be still used, will experience wear resulting in sharp edges which can cut into the liner. These edges or corners should be chamfered with a file when it is found necessary to relieve the scoring effect.

Liner Issues
It has been observed that if scoring of the liner for any cause is taking place, the metal very quickly finds its way into the packing and flushes out of the packing drain. If any discoloration of the oil or sign of metal particles takes place, the machine should be shut down and the liner checked for scoring through valve ports on each end of the cylinder. Replace the liner or implement the honing procedure described in Table 14.2.

The inspection of the packing drain oil need only consist of putting a small piece of tissue paper under the opened drain and examining the drop of oil under a light.

Many compressors have been converted to minimum lubrication ("minilube") or nonlube construction. Piston rider rings, also called wear bands, must be applied for this conversion. Occasionally, cast iron pistons have been too heavy for liners. Wear patterns here present themselves as holes or large pits in the piston. Rider rings can cure this problem. When rider rings are applied, check that the weight of piston divided by the projected area of rider ring is less than 8 psi (0.06 N/mm²) for minilube and 5 psi (0.03 N/mm²) for nonlube applications.

Cylinder Lubrication
Lubrication serves several purposes. It acts as a cooling agent, carrying away heat. It also washes away particulate matter and helps sealing. It prevents corrosion. However, its most important function is to reduce friction.[1]

Step 1: After the piston has been removed from the cylinder, install a container or some form of a catcher below the cylinder opening to prevent grindings from entering the working parts of compressor or crankcase.

Step 2: Thoroughly wash out cylinder with kerosene or an approved brand of commercial cleaning fluid; remove all deposits of carbon and oil, then dry with a clean linen rag.

Step 3: Determine the condition of cylinder to be honed. If the cylinder has been scored to depths of over 0.005″ (0.13 mm), a coarse stone "number 136" should be used to remove and dress down the extra build-up on the cylinder walls, leaving a depth of 0.002″ (0.05 mm) to be finished with stone "number 236." If there is no scoring present or scoring is less than 0.002" (0.05 mm), proceed to the deglazing step, Step no. 4.

Step 4: Deglazing—Use stone no. 236 or another of the 200 series; then follow the same procedure as in step no. 6.

All stones must be used dry on cast iron cylinders. Stones and guide blocks must be kept together and used as sets; never use the same pair of guide blocks with different sets of stones.

Hone driving motor. For best results it is recommended that a slow speed (250–400 RPM) air drill be used. The drill must have right-hand rotation and have a 3/8″ (9.5 mm) or larger chuck size.

Step 5: Insert and expand stones and guides firmly against cylinder walls by turning the winged collar on the hone clockwise. During this adjustment, stones should not extend more than ½″ (13 mm) out of the cylinder.

Step 6: Push the stone to the bottom of the cylinder; allow the stones to go through the lower end of the bore ½″ to 1″ (13–26 mm). Start stroking at the bottom of the cylinder using short strokes in order to concentrate honing in the smallest and important section of the cylinder. Gradually lengthen stroke as metal is removed and stones contact higher on the cylinder walls, stroke all the way to the top of the cylinder, always bearing in mind to maintain a constant steady stroke of about 30 cycles per minute. An excellent indication of cylinder condition is speed of the drill.

A reduction in drill speed, during honing, indicates a smaller diameter. Localize stroking at such sections until drill speed is constant over at least 75 percent of the cylinder length.

Step 7: Remove stone and check for cylinder appearance after free stroking, include no binding, for approximately 1 minute. The finished cylinder should indicate a diamond-shaped hatch pattern of approximately 25 to 30 μin (0.635 –0.762 μm) R_a* for the ideal seating of bronze piston rings, for instance. For Teflon rings, finish should be 16 to 20 μin. (0.41–0.51 μm) R_a.

Do not over-hone. If finish is too fine, ring wear will be excessive. Use roughness comparators to check the finish.

---

*R_a = RMS, or Root-Mean-Square

TABLE 14.2 Honing Procedure for Large Cast Iron Cylinders

Pressure (PSIG)	Cyl. Diameters	Minimum Viscosity (SSU @ 210°F)			
		0–10"	10"–15"	15"–20"	Above 20"
0–500		60–70	60–75	65–75	65–80
500–1000		70–80	70–85	75–85	75–90
1000–2000		80–100	85–100	—	—
2000–4000		100–150	—	—	—
4000 and Up		150–200	—	—	—

TABLE **14.3** Viscosities for Cylinder Lubrication (*Cooper Energy Services–Superior*)

Compressor lubrication oil must meet the following minimum requirements:*

1. Good wetting ability.
2. High film strength.
3. Good chemical stability.
4. Clean and well refined.
5. Oxidation and corrosion inhibitors not required, but may be beneficial.
6. Pour point must be equal to gas suction temperature plus 15°F to 20°F (7°C–10°C).
7. Good resistance to carbon deposits and sludge formation. If any carbon is formed, if should be the soft, loose, and flaky type.
8. Minimum flash point of 400°F (200°C).
9. For required viscosities follow Table 14.3.

Few mineral oils meet all of these requirements and expertly formulated synthetic lubes are best. As is so often the case, only a thorough experience check will give assurance of selecting the optimum cylinder lubricant. The cheapest lube is almost never the best from a cost-of-ownership point of view. Be sure to read our comment on page 152, taking issue with the "one-oil-fits-all" approach on critical process compression equipment. Protecting your assets is worth far more than the incremental cost of superior lubricants over the "conventional wisdom oil."

There is general agreement that determining the correct amount of proper lubrication for cylinder and packing is not an exact science.

*Cooper Energy Services (CES)—Superior.

The right amount of lubrication is dependent on the gas pressure, its variability, oil viscosity, and temperature. In order to determine required lube oil quantities considering these factors and cylinder sizes, compressor OEMs have developed formulas that provide an average lubrication usage for the normal application. Not only must the proper lubrication be applied in the proper amount, it must get to the proper place at the proper time. The little understood, much abused lubrication distribution, system shown in Fig. 14.3, is responsible for this task.

The distribution block type lubrication system in Fig. 14.3 uses positive displacement to meter the amount of oil fed forward to lubricate compressor cylinders and packing. Since the system operates on a proportional basis, a single adjustment at the force feed lubricator pump increases or decreases the flow proportional to every lubrication point served. The system consists of a pump, sized to provide enough lubrication and pressure to overcome the back pressure from the pressures inside the cylinder, and a series of valves, divider, or distribution blocks, to get the lubrication to the right locations. Hydraulically balanced pistons inside the blocks divide the oil into accurately metered amounts for each lubrication

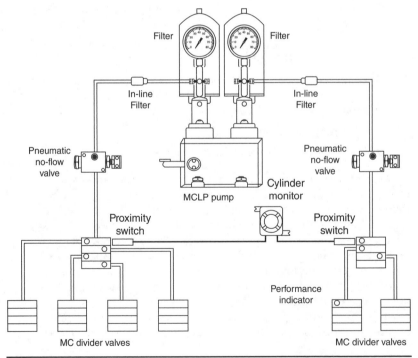

FIGURE 14.3 Divider block lubricator system. (*Source:* Lincoln Division of McNeil Corporation, St. Louis, MO)

FIGURE 14.4   Divider block. (*Source:* CPI Lubrication, Midland, TX)

point served. Sound selection and proper make-up of the distribution block allows

1. Accurately measured "shot" sizes, and
2. Precise proportioning to meet different or equal oil requirements.

A typical divider block, Fig. 14.4, provides a certain volume of lubricant with each stroke. For instance, a divider block of a given size delivers 0.20 in³ (3.28 cm³) per lubricator pump stroke. The compressor's per cylinder lubricant requirement value is met by making adjustments to the lubricator pump RPM, and total lubrication required allow to size a system that provides the proper amount of lubrication.

Because of the positive, metered operation, central warning equipment can sense trouble anywhere in the system. Safety equipment includes pin fault indicators in each outlet from the distribution block, a pneumatic or electric shutdown switch in the event of lubricant flow failures, and a rupture disc in the lubricator collector manifold.

It stands to reason that divider blocks have a finite life. Pumps can be adjusted to provide a minimum and a maximum amount of lubrication. If the lubrication system is sized for the mid-to-low range and the operator turns up the pump rate, more than the required amount of oil goes into the cylinder. When a divider block wears out it starts to leak. This leaking block can provide more lubrication than needed, or the oil may bypass the cylinder and not get to its intended destination.

Frequently, when a compressor is moved from one location to another, the pressures and volumes are different. It is possible that the cylinders were changed to meet the new conditions. However, quite often, the same divider block setup may end up being reused. Verify that it is still the right size for the new cylinder.

Lubricating oil consumption in pints per 24 hours is the figure on which cylinder lubrication is based. Use the number of drops viewed in a sight feed or similar observation port merely as a guide to obtain a balance between cylinder feeds. Recognize also that certain compressors may not be operating at their maximum RPM when setting the lubricator feeds by drops.

General formulas and reasonable approximations for calculating lubricant quantities for compressor cylinders are*

$$\text{Drops per minute} = \frac{\text{bore (in) stroke (in) RPM}}{10,000} \qquad (14.3)$$

$$\text{Gallons per 24 hours} = \frac{\text{bore (in) stroke (in) RPM}}{300,000} \qquad (14.4)$$

$$\text{in}^3/\text{day} = \frac{\text{bore (in) stroke (in) RPM}}{80} \qquad (14.5)$$

$$\text{cm}^3/\text{day} = \frac{\text{bore (in) stroke (in) RPM}}{1,300} \qquad (14.6)$$

Sight feed lubricators are often used; they facilitate checking the lubricant delivery rate. Manufacturers' specifications, for required lube oil quantities vary and are often contained in their operating and maintenance manuals. One high-speed, high average piston speed compressor OEM quotes a quantity of 0.2 pints/d/in of cylinder bore as being equivalent to one drop per minute per inch of bore for very heavy oils. The same OEM recommends quantities up to two drops per minute per inch of bore for very light oils.

To ensure that adequate lubrication is being achieved, periodic visual inspections of the cylinder bore and piston rod are recommended. Initial settings and adjustments to the feed rate should also be accomplished on a 24-hour basis because drop sizes vary with the viscosity of the oil.

It is interesting to note that even the manufacturers of Teflon are tying their product to a cylinder lubrication issue. They advise that

---

*Generally accepted: 10,000 to 14,500 drops per pint (~473 cm³) depending on viscosity.

their product is only truly self-lubricating when it is rubbing on Teflon. This means that metal coming into rubbing contact with Teflon must pick up enough of this polymer to form a very thin coating over the metal surface. In fact, rapid wear of Teflon will occur until this coating is formed and new or rehoned liners will experience accelerated initial wear on the replacement Teflon rings. This wear is self-limiting and will be followed by normal rates of wear. It is especially noticeable on nonlubricated (Teflon-equipped) reciprocating compressors.

Therefore, new or rehoned cylinders should be taken out of service and the rings measured at frequent intervals until the first set of rings indicate their wear-out time.

Ring Thickness
Plotting ring thickness on a time graph should show a rapidly decreasing rate of wear and the life limit can be predicted. This plot should be followed on all new liners. Honing newly bored liners with Teflon blocks will reduce this rapid wear somewhat if done after preservative grease is removed.

Dirt
Impurities or contaminants, usually present in the form of dust and liquid carryover, are the enemy of compressor cylinders. Lubrication in the cylinders picks up the dust and makes an abrasive sludge which damages liners, rods, valves, rings, and packing. Nonlube machines transport dust easier with less wear, but in either case dirt should be prevented. Liquids wash the lubrication off cylinder liners and piston rods and cause scoring; small amounts of liquid cause minor valve abrasion. Large slugs of liquid can instantly destroy valves. Liquids on valve seats can flash and cut the metal during normal operation. While it can drastically shorten valve life, it can also cause cavitation-erosion on the seat. Damaged valve seats invite blow-by and valve overheating.

Suction Scrubbers
Several types of suction filters and liquid disengagement vessels have been tried with varying degrees of success. The best type found so far is the two-compartment type with filter cores for solids removal and liquids drop-out in the first compartment. Demister-mesh or crinkled wire mesh (CWM) is accommodated in the second compartment. As mentioned in an earlier chapter, every attempt should be made to obtain the best-possible filter, knock-out drum, or filter-coalescer just upstream of the compressor suction nozzle. Good manufacturers and owner-purchasers recognize that liquid disengagement will occur orders of magnitude more effectively at low gas velocities. For a given volume of throughput, a larger contact area will keep gas velocities down. Larger area of contact mean large vessel, which is more money

up front. But it will lower the cost of ownership because there is greater availability and fewer catastrophic downtime events with well-designed, larger vessels.

Beware of piping inserted between any such disengagement vessel and the compressor suction nozzle. If the gas is at or near saturation conditions, cooling this exposed suction piping will cause vapors to condense and enter the compressor. Think of the eventuality of rainwater hitting it. Will condensation then occur?

There is a continuing effort by compressor knockout filter supplier to improve their products. Regular cleanout of the filter must be scheduled.

Packing Care

Packing Rings
As mentioned in a previous chapter, *segmental ring packing* is a standard component used to seal the space between the compressor cylinder and the piston rod. Teflon is used whenever practicable. Packing rings rub on the hardened area of the piston rod and heat is built up, which is additive to the compression heat in the cylinder. Keeping the packing and rod cool is therefore, a first requirement for good care.

Packing cases (or glands) for higher compression ratios or high gas temperatures often have cooling passages. This is one area where the coolest water available is best. Since the segments are stacked, care must be taken to line up and seal the cooling passages. In addition to cooling, most packing is lubricated by injection of oil into the innermost cup and sometimes again in other cups toward the outside, depending on the number of cups in the packing gland. A drain and vent is also located in or near the last cup and sometimes on deeper cups. Each vent and drain line is separately piped out of the gland. These drains give an excellent indication, not only of the condition of the packing, but also of the cylinder itself. The vents should be regularly sampled for condition of the oil drain and blow-by; see Fig. 14.5 later.

Frequently, packing vents are instrumented for flow indication, and a Rotameter or similar device can be used as a means of monitoring machinery health. Observing oil drained from the sample valve, Fig. 14.5, is important. If it has turned dark or starts containing particles, scoring of the liner is probably in progress. Oil that is emulsified and yellowish indicates that the packing is not seated well or may be damaged. Corrective action should be taken as soon as these defects are noticed. The same indicators are present in nonlube service but are much more difficult to see. Still—if in doubt, it is better to find and remedy a problem early rather than wait for a major failure. To say that human lives are at stake is an understatement.

Dirt carryover will foul the packing cups. If vents and drains show darkened oil, the rod will usually exhibit scratching: Clean the packing as soon as possible. The, usually combined, vent and drain

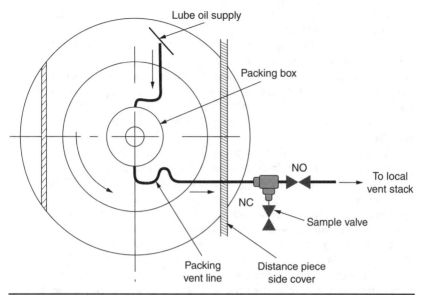

Figure 14.5   Packing lubrication and vent connections.

line is the best indicator of cylinder and packing scoring. Keeping a regular watch on it and taking preventive action without delay can keep the machine in operation for long periods. Daily taking a few drops of oil from the packing vent will serve as a relatively quick indicator of an approaching need to shut down the compressor.

Breaking-in Procedure

Breaking in new rod packing, particularly in older rods, is a tedious task. The most important task is to keep the rod cool. Set oil rate about three times of normal levels but not at full flow since over-lubrication for break-in will cause the cups to fill and have a hydraulic pressurization effect on the packing. Also, carbon and sludge can be formed as oil overheats. Keep the inspection covers off the distance piece and use only enough oil to see a reasonable drip from the drains. Add oil to the rod in the distance piece by hand to keep it cool. For nonlube service with Teflon packing, apply a reasonably small amount of water from an oil can to keep the rod wet and cooled. The packing should seat in about 2 hours and thereafter give acceptable blow-by.

Friction is the major cause of heat buildup in a packing case. Under normal circumstances, lubrication on a piston rod as it moves through the packing should be evenly applied; it should result in an oil film thickness of about 0.002 in (51 μm). Such a lubricating film lowers the friction factor and extends ring wear and life. In many cases Eqs. 14.3 to 14.6 are also used to determine lubrication requirements for the piston rod and packing by treating it as a separate cylinder. However, as average piston speeds increase, over-lubrication will become a problem.

As a preventive maintenance (PM) check, experienced maintenance technicians sometimes use cigarette paper to determine if there was enough or too much lubrication. The same approach is described in several OEM service manuals. Essentially, the technique uses two sheets of paper to wipe down the cylinder bore or rod. To signal acceptable lubrication, the first sheet should be oily and the second one should be dry.

Another experienced authority believes a single sheet of cigarette paper can be used as an indicator. Over-lubrication is certain if, after patting the paper into the cylinder walls and pulling it out, the oil not only covers the entire paper but droplets drip from it. Finding only isolated or discontinuous spots of oil on the paper would mean insufficient lubrication. We believe either method is reasonable.

Over-lubrication will prevent the packing from sealing or cause it to fail within a short period of time after startup. Smaller, high average-piston-speed compressors are more susceptible to over-lubrication than larger, slower compressors. There are two different scenarios: One involves too much lubrication in the packing case and on the rod when installing new packing. The other is a result of continuing to add too much lubrication through the lubrication system until it stops the packing from functioning. One should suspect over-lubrication when the packing will not seat on start-up. While there are other factors that can contribute to a failure to seat, over-lubrication is by far the most prevalent.

If the maintenance technician replaces the rings with another set, he or she should examine the inner bore for bad rings. If over-lubrication was the reason for their failure to seat, most likely they will show no or very little wear marks from rubbing on the rod. This is because the rings never touched the rod. They were probably riding on top of the thick oil film; by not actually touching the rod surface, the rings failed to perform their intended sealing function.

In applications where the rings have sealed initially but fail after a week or two, the small diametral clearance of ~0.004 in (~0.1 mm) provided for the rings to move, adjust, and distribute the loads evenly. The rings are thought to have later filled with excessive amounts of oil and, often, with other liquids. This somehow rendered the rings unable to move; they then overheated, partially melted, and failed. When removed for examination, these rings are often found saturated with oil and stuck together, which is why the above explanation seems quite plausible.

In the first case, where rings will not seat, consider that carbon-filled Teflon rings do not require much lubrication. They were originally designed to run in nonlube applications. Thousands of these rings run in nonlubricated applications in the petrochemical industry.

Their useful operating life may not always match that of high-performance polymers, but they work well. Unless there is a lot of washing, as mentioned before, a very thin film of lubrication is sufficient.

When packing will not seat on start-up, wiping the rod down with a rag often helps. Rotate the crank shaft so the rod travels in and out of the packing case a couple of times and wipe it each time. This will help remove the excess lubrication from the rod and case.

In the second case, where rings fail after a relatively short operating time, determine the lubricant consumption and check for too much lubrication. Always have a qualified individual determine required lubrication rates.

Here are the steps to follow if over-lubrication problems are suspected:

1. Determine the cycle time of the divider block used with the application point. With a stopwatch or the seconds-hand of a watch, measure the time for one complete cycle of the pin or LED flash. Record this time.

2. Calculate the proper lubrication rates.

For a two-throw high-speed compressor with an average piston speed of 900 fpm (4.6 m/s) at 900 RPM referred to on page 148:

$$2 \text{ in packing} = 2 \times 2/5 \text{ pint/d} = 0.8 \text{ pint/d} \ (4\text{--}8 \text{ drops/min})$$

$$2.5 \text{ in packing} = 2.5 \times 2/5 \text{ pint/d} = 1 \text{ pint/d} \ (5\text{--}10 \text{ drops/min})$$

Lube Application Rate

Changing lubrication amounts or rates is often appropriate. However, before attempting to reduce the lube rate, have a competent individual size the blocks and determine if the lubrication rate can be reduced. Remember that lowering the pump rate lowers the lubrication rate to each part of the system. Possibly, the problem may just be a case of oversized divider blocks. Divider blocks are relatively inexpensive and easy to change.

If the lube rate needs to be reduced, use the manufacturer's procedures to lower the lubrication. Only reduce lubrication flow by about 5% at a time. Allow the compressor to run a week or two, then inspect for excessive ring wear before lowering the rate once again.

Scoring

When the rod begins to score and shows indentations of 1/16 inches (~1.5 mm) or deeper, using new packing is unlikely to solve blow-by problems. Expect packing life to be very short with severely scored rods. Rods are very expensive and the practice is to re-plate them.

The important feature of plating is to obtain a hard sliding surface with enough porosity to hold lubrication. Alternatively, a plated surface rubbing against Teflon must pick up and hold a mating layer of Teflon. Mirror finishes will not permit lube oil to stick and adhere; a measure of surface porosity is needed to hold the lubricant. Without lubricant, the rod surface will overheat and spider web cracks will form. After a while, pieces will flake off and chew up the packing. Chrome must be closely watched for this fault. Recommended rod finishes are 8 to 12 μin (0.20–0.30 μm) R_a on Teflon and 95 μin (2.45 μm) R_a maximum on bronze.

Rod Coatings

The economics of rod coatings are always important. Chrome is cheapest but has occasional problems of flaking off. Colmonoy is better than chrome but is not always readily available; it requires skill to apply it. Tungsten carbide and chromium carbide coatings will cost about the same as a new rod. However, with proper care, either of these coatings will last the life of the machine.

Valve Care

A valve plate lifts as the piston is approaching its maximum speed, usually one corner first, then the valve plate rises until it strikes the guard. It then returns to the seat at the end of the stroke, usually by spring pressure. It is then held in place by the pressure across the valve. If the valve plate strikes the guard hard enough to rebound to the seat, the result is valve flutter, a condition that can destroy a valve in a matter of hours. Valves today (2012) are all designed with this problem in mind so this problem occurs rarely.

Compressor unloaders were ranked one of top six causes of unscheduled compressor shutdowns.* A number of compressor maintenance surveys confirmed that unloader-related shutdowns can be associated with, although they are not limited to:

(a) Leakage of process gas (fugitive emissions) through the unloader packing at levels higher than allowed by applicable emissions and/or safety regulations

(b) Unloader seat leakage which can result from improper centering/alignment of the unloader in the cylinder

(c) Unloader sticking/seizure preventing free movement of the unloader

(d) Unloader vibration resulting from an excessive pressure drop across the unloader valve

---

*Reference 2.

FIGURE 14.6 Valve guard with springs and damper plate.

One way to reduce maintenance concerns associated with unloaders is to simply eliminate those capacity load steps that are no longer required. Removal of unnecessary unloaders simplifies the compression system and eliminates a possible source of maintenance. Upgrading the remaining unloaders can significantly reduce or even eliminate some of the concerns outlined above.*

Valve Flutter

Valve flutter is frequently mentioned as a valve problem; however, present-day valve design and good application have largely eliminated valve flutter. One valve manufacturer has designed a second plate behind the main valve plate with a heavy set of springs; the second plate will cushion the valve plate travel and mitigate flutter, see Fig. 14.6. Other manufacturers may use additional springs or gas traps to cushion plate impact.

Valve Lift

As mentioned in an earlier chapter, total valve lift is very important with respect to efficiency and therefore power consumption. Total valve lift is also important in preventing flutter. Generally speaking, gas with heavier molecular weight (molar mass 20 and above) would have a lift of 0.08 in (2 mm) and gases with molecular weights of 7 and below would have approximately 0.040 in (1 mm) lift, with

---

*Dresser Rand Literature—Ref. 2.

some proportional lift between these values. Equations 14.7 and 14.8 explain gas velocities in the actual valve:

$$V = 288 \frac{D}{A} \qquad (14.7)$$

or

$$W = \frac{Fc_m}{f} \qquad (14.8)$$

where V = average gas velocity, in feet per minute (ft/min)
 D = piston displacement per cylinder, in actual cubic feet per minute (ft³/min)
 A = total valve area of all suction valves or the product of the actual valve lift, the valve-opening periphery, and number of inlet or discharge valves per cylinder, in square inches (in²)
 W = average gas velocity, in meters per second (m/s)
 F = piston area, the area of the crank-end of the cylinder less the piston rod plus the area of the outer end of the piston, in square centimeters (cm²)
 f = product of the actual valve lift, the valve-opening periphery, and the number of inlet valves per cylinder, in square centimeters (cm²)
 c_m = average piston speed, in meters per second (m/s)

API-Recommended Velocities

These are larger than those normally used by industry. The compressor manufacturers, in their concern for low valve horsepower losses, have often used larger valves (lower API velocities) than the optimum needed for stable valve action. As an example, one manufacturer recommends API velocities for air of about 3200 fpm (16 m/s), and for hydrogen of about 7500 ft/min (38 m/s) at 15 psia (103 kPa absolute) and 3500 ft/min (18 m/s) at 200 psia (1380 kPa absolute) suction pressure.

Valve Plate Materials

As of the issue date of this text, most owner-operators had managed to upgrade their valve plates from metallic to thermoplastic materials. Frequently however, they do not know or control the exact composition of the material they are using. They express surprise when valve failures are experienced due to higher than normal operating temperatures. Such failure occasions represent an opportunity to extend the temperature range of valve rings by choosing a different thermoplastic.

Valve Springs

These serve to control the timing of valve closings. The springs are subjected to dynamic loading by the motion of the valve element. Element motion is periodic, but includes rapid acceleration and deceleration. It can therefore excite a wide range of frequencies in the spring. The resulting spring surge creates high stresses and is a major contributor to premature spring failure.

Valve failures are, as noted earlier, the most common reason for unscheduled compressor shutdowns. Valve springs in turn are the most frequent cause of valve failures. Therefore, valve spring life is of utmost importance in attaining increased compressor reliability.[*] If valve failures are caused by corrosion, specifying PEEK[†] valve plates with INCONEL X-750 springs, and piston rods of 17-7 PH stainless steel will very often eliminate this failure mode.

Checking Valve Lifts

An easy method of checking valve lifts and checking whether or not the valve assembly has free movement or is sticking is as follows: When the valve is assembled, a depth micrometer is used in an outlet port. Push through the outlet port and activate the plate through its full travel. The differential depth from when the plate is opened versus when it is closed will yield the valve lift. Generally, this requires more than one reading to determine if there is uneven wear of the stop plate. If there are strong spring forces, the valve plate may have to be pushed at two or more points simultaneously to guarantee that the plate has bottomed flatly onto the valve stop.

The best way to check valve lift is to disassemble the valve and measure its dimensions at multiple locations. Then, stacking up of dimensions will allow us to determine if the valve lift is correct. At disassembly and while measuring, we can perform a visual examination. Look for signs or contours of wear, mark the thickness of the plate, observe high spots, side clearances, etc.

When a valve is removed for checking and the plates are being replaced, care should be taken that both seat and guard are checked for wear. Lap both if in doubt.

Examine the replacement plates and seats, and also the guards for any sharp edges or corners. Since the plates lift off at a corner first and not straight up, any sharp edge or corner will give short valve life.

Suction valve unloaders are often a source of trouble. One ought to be wary of unloaders that lift a whole valve off the seat. Sooner or later the assembly will cock and break. Watch the fingers of suction valve lifters. They can crack when they strike the port when lifting the cover and then later break off in service. Work them by hand every time they are out for inspection.

Fastener Torque

Never over-torque valve cover nuts or jackscrews. This can distort the valve and lead to valve cover failures. If the cover leaks use a new

---

[*]See Ref. 2.
[†]Polyether-ether-ketone (PEEK GF30); Hoerbiger HP.

gasket and if that does not work, lap the cover on its seat. Radial lines indicate a leak or the start of a leak. If radial lines are visible on the gasket seat, the seat should be lapped.

Air compressors with their higher compression ratios and, therefore, higher adiabatic discharge temperatures, often have trouble with lube oil breakdown. The oil forms a carbon-based varnish which then coats the valve and seat. The coating action can cause valve plate distortion and early breakage. For these applications, take special note that discharge temperatures stay below 375°F (190°C) and be certain that only the best lubricant is used.

Liquid carryover is another valve problem. The liquid sits on the seat; it tends to distort the valve and often flashes to vapor between valve and seat. This erodes ("cuts") the metal and will cause harm to valve seats. Whenever erosion is noted on valve seats, process and equipment design should be questioned as to the adequacy of knock-out facilities in the system. (See index for key words *liquid disengagement or knockout drums*.)

Valves in nonlubricated service tend to experience more problems than valves in services with cylinder lubrication. Manufacturers are utilizing Teflon or PEEK and other similar material components and coatings which have had success on process machines where temperatures have been held below 350°F (177°C). In many new applications, however, trial and error for valve selection is still a proven method.

Temperature is the best indicator of valve malfunction. The old-fashioned method of carefully and briefly touching each valve cover and noting if any one cover seems warmer than the others still remains a valid check. Whenever temperature deviations are suspected, and certainly if temperatures are over 150°F (65°C), hand-held pyrometers, often contact probes, should be used. Ultrasonic inspection using a hand-held probe or aiming a similar instrument at each valve cover has also been successful, as have thermographic (thermal monitoring) surveys. Both methods are very application and machine operation specific; either should be developed on a per-site basis with expert involvement.

Valve Repair Practices

Maintenance and repair procedures differ among the various valve types. However, certain general practices are shared by many compressor valve configurations.

A guideline for valve inspection frequency cannot be given; it clearly depends on the type of compressor, the operating conditions, and properties of gas being compressed. Valve inspection and repair strategies vary from periodic inspection routines to wholesale valve replacement based on accumulated run time. Reference 3* points to

*Pages 479 to 497.

statistical methods using Weibull analysis. A renewal function can be calculated by first determining the characteristic life and the failure patterns of compressor valves at one's plant. Data collection is critically important to Weibull analysis and owner-operators have been able to arrive at the optimum, lowest cost complete set replacement interval by feeding data into commercially available Weibull software. In new compressor installations, compressor valves should be inspected and cleaned within the first 1000 hours. At that time, slight wear on valve plate and valve seat might give an indication of desirable maintenance schedules.

When valves are only inspected and cleaned and no parts are remachined or replaced, matching the same valve seat and valve plate as originally assembled is needed. This is necessary to ensure proper tightness since the seating pattern will match perfectly.

Valve Component Replacement
When to replace valve components at inspection is a valid question. As a general rule, if the wear on the valve plate is even and less than 10% of the original plate thickness, the plate can be cleaned and reused. Valve plates that are worn on one side should never be used in the reverse position. If the valve plate shows more wear than stated earlier or uneven wear or pitting, replace it along with all springs and machine, grind or lap the seat. If the valve seat is remachined, the valve plate and all springs should be replaced.

Valve Component Refurbishing
A valve seat is shown in Fig. 14.7. The seat face should be lapped every time a valve plate is replaced to prevent leakage, because the seating pattern from one valve plate to another is always different. Remachining of the seat face is almost always necessary when a metal valve plate has broken. There will be uneven seat wear at the location of breakage.

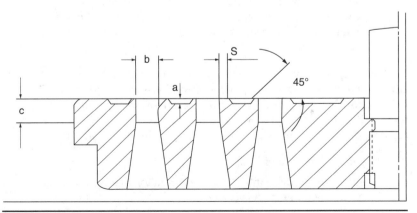

FIGURE **14.7** Valve seat.

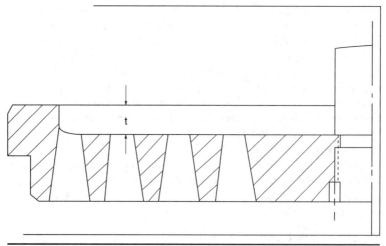

Figure 14.8   Valve guard.

If the valve seat surface shows normal wear and no damage to the seat lands is obvious (dimension "S"), lapping or concentric grinding is sufficient. The seat band width "S" should be maintained to original dimensions when remachining the seat surface. A valve seat can be remachined without affecting the valve performance until the ratio of b-to-c becomes 1:1 on drilled seats, and 3:2 on milled or cast seats. The bottom of the groove should not be machined to maintain the proper b-to-c ratio; this would weaken the seat.

- *Valve guards* (see Fig. 14.8). Valve guards do not normally have to be re-machined. If, for any reason, the top side is re-machined, dimension "t" should be at least equal to the guide ring thickness (if applicable).

- *Valve plates.* Metal valve plates should not be reground. If the wear on the plate exceeds 10% of the plate thickness, it should be replaced. Coil springs should be replaced every time valves are overhauled.

Valve repairs have to be carefully managed. They should be executed in a dedicated valve shop where cleanliness would be of utmost importance. Many reciprocating compressor owner-operators send their valves to a reputable valve repair facility. The chosen valve repair shop should adhere to the following minimum requirements:

1. Each valve is identified.
2. Each valve is dismantled and inspected for wear or breakage. An inspection report is issued.
3. Seats, guards, bolts, etc. are cleaned with a vapor blaster.

4. Seats and guards are inspected for cracks.

5. Seats and guards are re-machined to OEM specifications and, if necessary, metalized before machining. Seats are lapped or concentrically ground.

6. Each valve is assembled. Only OEM parts are used.

7. Each valve is function tested and two different leakage tests are required: (a) liquid test (1.5 minutes and slots of seat must be still half full) and (b) air test (using a flowmeter).

8. Each valve (if not nonlube) is dipped into rust preventing oil and wrapped.

9. Each valve is identified by a sticker on the wrapping.

Materials of construction for piston rings, wear bands, packing, and valve plates. It is very important that the materials of construction (MOC) of rider rings (wear bands), compression rings, packing rings, and valve plates are clearly identified by the compressor purchasers and owner-operators. Once this information is available, there is an opportunity for stepwise life improvement of these components. This improvement should go hand-in-hand with a well-established computerized maintenance management system (CMMS); a CMMS must capture life data on all compressor components.

If piston and packing ring and also valve ring life are unsatisfactory, a controlled program of introducing new materials should be considered. The first thing to do in a component life improvement program should be, again, the identification of mating materials such as cylinder wall and piston rod surfaces. These materials might have to be upgraded by a proper finish for the cylinder wall or by suitable coatings* applied to the piston rod with a finish ultimately to be defined by the packing ring supplier's engineering department.

The most common materials for nonlubricated but also lubricated applications in reciprocating compressors are PEEK (polyether-etherketone), PPS (polyphenylensulfide), PI (polyimide), and PAI (polyamideimide). These "high performance" thermoplastics, in their pure form, do not have the ability to run dry. However, they become very resistant to wear once they are modified by additions of such lubricating substances as PTFE (10%–15% by weight), graphite, and molybdenum disulfide.

If the compressor OEM cannot assist in extending the life of the various wear components, owner-operators must work with well-established compressor product companies. These companies must be

---

*The portion of the piston rod which operates in the packing should be coated with Praxair's Detonation Thermal Gun, "D Gun" (or equal), plasma-sprayed Tungsten Carbide with a thickness of 0.005 in (0.12 mm). Where filled PTFE packing is used, the surface finish of this coating should be 3 to 5 μin RMS.

engaged in producing piston and rider rings, piston rod packing, and valve components for reliability-focused users. The favored suppliers may not be lowest in first cost, but will be superior to the competition based on failure risk reduction and safe operations strategies.

Noise and Vibration

Every reciprocating machine develops its own basic operating sound. Good operators can tune their ears to this sound and recognize changes during operation. Should changing sound patterns emerge, the facility must investigate the cause. It has been said that using noise as an indicator triggering preventive maintenance makes much economic sense. Of course, other predictive maintenance methods must also be applied in a reliability-conscious plant environment.

Knocks are the most common abnormal noise. Finding the cause of the knock is often more difficult than hearing it. Oddly enough, the first place to look for is in the cylinder adjacent to the crank throw where the knock is located. This is because, in almost every instance, a knock in the cylinder will be heard in the crankcase before it can be heard in the cylinder. The compressor piston nut could have come loose, or the compressor rod could unscrew from the crosshead. A knock can be heard from the piston sliding on the rod. If the piston or nut strikes the cylinder end, the knock will be quite pronounced.

Bullet-nosed pistons can droop due to excessive wear and the fillet radius on the piston end can strike the fillet radius on the cylinder. Broken valve plates drop into the cylinder and can interfere at the end of the stroke. Any of these faults can be readily located by cylinder inspection through a valve port. They can, however, cause a machine wreck rather quickly, which is why a precautionary machine shutdown or instant trip should be initiated as soon as knocking is heard and located.

Sometimes the compressor rod nut works loose repeatedly after re-torquing. Cylinder misalignment is the most probable cause. This can be rooted in broken grout under the crosshead pedestal, or by wedging up too tight under the discharge bottle. A fast check is to place a machinist level on the crosshead slipper and on the cylinder bore through a valve port, while the machine is still hot. If the bubbles in the levels are not identically placed in the glass, after instruments have been checked together side by side, then there is cause enough to examine the distance piece and compressor flanges for broken studs, or the unit may have to be realigned and regrouted. Always ascertain that discharge bottle wedges are not too tight after the machine reaches stable operating temperature.

Another cause of cylinder knocks is too tight a bump clearance on the head end. Every time the nut or rod has been backed off, the bump clearance must be taken and is to be set with the end clearance allowance as recommended by the manufacturer allowing for rod expansion when heated. Frequently this information is shown on the cylinder nameplate. The rule-of-thumb is one-third of total clearance on the crank end and two-third on the head end of a cylinder.

Modern heavy duty compressors are built sturdy enough, so problems with crank cases occur rarely. Crosshead bushings, crosshead shoes, connection rod shells and main bearings, with proper lubrication, should last the life of a machine, generally assumed to be at least 20 years (although some reciprocating compressors are still in service 60 years after commissioning). If knocks occur but cannot be located in the cylinder, the wrist pin clearances should be checked. Feeler gauges in the upper crosshead shoe can check for crosshead slap. Shake the connecting rod (the "conrod") and feel for indications of slack shell side clearance. Jack up the crankshaft at the main bearings (the "mains") against a dial indicator to determine main bearing wear.

Frame (Crankcase) Repair Practices

Foundations

Reciprocating forces originating in the compressor frame (sometimes labeled the crankcase) must be absorbed or dissipated by passing through a concrete foundation and into the soil. For this to happen there must be a firm attachment between the frame and the foundation. A cement or epoxy-based grout material must serve as the bonding material between the two. Unfortunately and over time, leakage of oil tends to degrade cementitious grouts; this can cause the frame to break free of the grout and start sliding back and forth relative to the foundation. This can be seen by examining the gap (usually oil-filled) between the grout and the side of the frame. Any "breathing" seen at that location is a danger signal. The deflection of the frame or crankcase will be passed on and result in similar deflection of the crankshaft—a form of reverse-bending under stress. This can soon cause a broken crankshaft and a probable machine wreck.

Tightening the anchor bolts will not help; they are not designed to absorb the shaking forces acting on good foundations. Fortunately, effective and fast in-place pressure grouting systems have been developed. In case an entire foundation is in poor condition, the machine should be jacked up on railroad ties while the concrete is being refaced and cracks filled.

Some foundation designs are troublesome because they leave only 1 to 2 in (25–50 mm) clearance under the oil pan of the frame. There are even less acceptable designs that leave no space at all. This causes the foundation to grow in the center from heat being generated from the crankcase and tilts the region under the frame flange. Broken

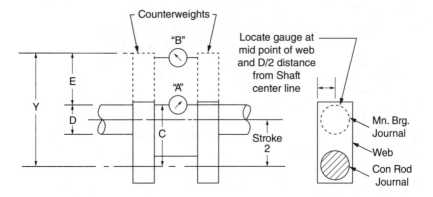

Zero (starting) position of crank
Formula for converting "B" to "A" − A = B reading × C ÷ Y where "C" = 1/2 stroke + D/2.
Y = C + E

Figure 14.9   Crankshaft web deflection nomenclature and measurement locations.

anchor bolts are a likely result. To prevent this from happening, particularly when regrouting, leave a 12 in (30 cm) space under the pan. Leave the ends open to allow for ventilation, then place grout under the crankcase frame flange only. This, after all, is the only region where grout is required. The pan clearance is needed when the unit is mounted on sole plates or rails that are grouted in.

Crankshaft Deflections

The crankshaft is the heaviest and most expensive component of a reciprocating machine. It incorporates throws which can bend, fatigue, and break. If there is bending it will inevitably occur between the throws, which means that measuring here is of greatest importance.

Crank shaft web deflections can be checked by utilizing electronic deflection indicators.* If such an instrument is not available, use dial indicators. Dial indicator measurements are done by placing a dial gauge between the outer ends of the crank webs as shown in Fig. 14.9. Turn the crankshaft to place this crank at bottom dead center position, 0°. Set the dial gauge at zero (0). With the gauge thus set and in place, rotate the crankshaft and take readings of the gauge at the back quarter position (90°), top center position (180°), and front quarter position of the crankpin (270°).

Where the connecting rod interferes with the indicator it will be necessary to move it out of the way. Calculate the reading back to the main journal position as shown in Fig. 14.10. Complete a table showing

---

*See Ref. 1.

Crankshaft Deflections						Machine No. _____		
Throw	1	2	3	4	5	6	7	8
0°								
90°								
180°								
270°								
Crankcase Temp. _____					Date _____			

FIGURE **14.10**   Crankshaft web deflection readings.

deflections and temperatures. Keep the data in the compressor file for periodic future comparisons—trending incipient defects. Deflections change with temperature and hot deflections are considered accurate and relevant readings.

One throw showing an inconsistent reading at 180° is an indication of a wiped main bearing. A wiped bearing will also show deflection carryovers to the 90° and 270° position. Address the situation by rolling out the bearing shell and have it re-babbitted or replaced.

Realize that there are different grades of babbitt, some hard, some softer. The harder babbitt alloys have greater load-carrying capacities than the softer grades. But the harder babbitts will not be as (relatively) tolerant of dirty oil as will as the softer grades. Everything is a trade-off; try to be guided by competent manufacturers or non-OEM specialist rebuilders.

If the deflections show an increase in negative readings on the dial toward the center, take it as an indication that the crankcase has come free of the grout and is deflected down on both ends. This is usually caused by the ends of the machine sliding on the grout and wearing it down. Only re-grouting will fix this problem.

Deflections showing an increase in positive readings to the center should be viewed as foundation sag. Again, the foundation should be fixed.

Readings that show a change in measurement, positive or negative, at any position indicate an S-bend; this would probably be caused by worn main bearings across the machine. This can be confirmed by taking bearing lifts by putting a jack under the crank at each main in succession and locate an indicator on the top of the crank. Jacking up the crank will give the clearance in each main.

Crankshaft deflections are solid indicators of the state of health of the crankcase. Again, these readings should be taken periodically based on operating hours, i.e., once per year or so. If the deflections show no increase between inspection events, the crankcase is staying in good repair and no problem should be expected. If changes are noticed, it would be very wise to perform annually a simple nonintrusive visual inspection. Further hints on compressor asset management are beyond the scope of this text, but can be found in Ref. 4.

What We Have Learned

- Repair practices must be predefined and discussed with service providers and contractors.

- Dimensional limits (condemnation limits) on compressor valves and other applicable components must be agreed to by all involved parties.

- Repair procedures should be developed and followed by personnel involved in the repair process both on-site and off-site.

- To be profitable, an owner-user company will have to groom its own reciprocating compressor specialists. Grooming means mentoring, tutoring, teaching, learning, implementing with forethought and consistency. These pursuits take time and will be well worth the investment. Retain your compressor specialists by giving them absolute and unswervingly equitable pay or salary treatment.

- On reciprocating compressors, be certain to never allow several deviations to stack up. Understand the consequences and communicate your concerns to higher management.

- Buying reciprocating compressors of sturdy, conservative design will always be a wise investment.

References

1. Bloch, Heinz P., and John Hoefner, *Reciprocating Compressors: Operation and Maintenance*, Gulf Publishing Company, Houston, TX, 1996.
2. Bloch, Heinz P., and Fred Geitner, *Machinery Failure Analysis and Troubleshooting*, 4th ed, Elsevier Publishing Company, London, UK, 2012.
3. Bloch, Heinz P., and Fred Geitner, *Machinery Component Maintenance and Repair*, 3d ed., Gulf Publishing Company, Houston, TX, 2004.
4. Bloch, Heinz P., and Fred Geitner, *Major Process Equipment Maintenance and Repair*, 2d ed., Gulf Publishing Company, Houston, TX, 1996.

CHAPTER 15

Maintenance and Operations Interfaces

Close cooperation between maintenance and operations personnel is an absolute requirement to ensure reliability and availability of reciprocating compressors. Operating and maintenance technicians must make a certain number of joint decisions well in advance. These two functional groups must also reach a firm understanding on who will be responsible for data taking on compression equipment. Over the decades, the best and most profitable facilities have opted to assign data collection to the operators. Corrective action steps requiring tools are assigned to the maintenance technicians.

At facilities with multitask training, the data taking and corrective action assignments often reside within the same group, but a number of limitations or boundaries are always spelled out and predefined.

After reaching consensus on the following, a major overhaul checklist (see end of chapter) should be compiled.

Shutdowns

Shutdown switches can be a nuisance—use only when really justified. The few essential "shutdown triggers" are:

(a) High gas discharge temperature for each cylinder

(b) Low-frame lube oil pressure

(c) High-frame vibration

(d) High level in upstream moisture separator or suction scrubber

(e) Driver overspeed, if other than an electric motor

The compressor and driver should be shut down by each of these switches at least once per year to check for proper operation; if found faulty, the situation should be cleared up or the device replaced. These shutdowns should never be blocked out, bypassed, or ignored. Doing so will imperil physical assets and human lives. The time to test some of these devices is quite literally one minute before the scheduled shutdown commences. Also, operators should be assigned well-specified tasks to be handled during the shutdown. It is rather obvious that the various departments that operate, maintain, trouble-shoot and analyze reciprocating compressors must cooperate very closely.

Monitoring and Record Keeping

Keeping logs from a maintenance standpoint is to determine the wear rate of the machine, to plan or schedule maintenance shutdowns, and to determine as closely as possible what must be done to minimize downtime. Plotting some of these readings on a curve with a time base can be helpful to recognize trends. Here are some suggested readings and checks:

(a) Daily
   Lube oil pressure, deviation from normal range
   Lube oil filter delta P (Δp, pressure drop)
   Compressor discharge temperature and pressure (deviation)
   Compressor packing vent, evidence of contaminants in leakage
   Lube oil consumption, deviation from range deemed acceptable

(b) Monthly
   Lube consumption, crankcase
   Lube consumption, cylinder lubricator
   Operating hours
   Compressor rod drop (if not already automatically provided by permanently installed rod drop-monitoring devices)
   Safety switches (emergency shutdown or ESD tests)

(c) Annually for first operating year—then increase interval as experience is gained
   Crank deflection
   Compressor valves
   Compressor rings
   Compressor packing
   Rod and liner finish
   Connecting rod clearance
   Wrist pin clearance
   Crosshead clearance

Recording the scope, nature, and results of repairs on a specific valve, packing gland, or ring is essential. Generalized terms (such as "replaced bearings") are not very useful for the reliability professional whose task is to prevent future events. The record must reflect details as to why, when, and where maintenance or remedial action was performed.

Cleanliness and Tools

Operators have many tasks that involve blinds being inserted in piping, manipulating valves, observing component condition, tracking progress, updating procedures, taking photographs, etc. Make them the guardians of cleanliness and the watchdogs of proper tool usage. Their lives will be on the line later, so make them your deputy inspectors and quality control enforcers.

Cleanliness is vital in machinery. Seemingly harmless dirt gathers in some places inside a compressor during operation. This accumulation then becomes disturbed when the machine is opened. Therefore, open machines up only when there is a good reason to do so. Before closing it up, wipe it down carefully and thoroughly.

When closing up a reciprocating compressor, special care must be taken that dowels fit properly; experience shows that they often hang up on burrs. Torque all nuts correctly. Never over-torque nuts because doing so distorts castings, bearing, and valve caps. Distortion will cause early failure. Make certain that the piston rod nuts are fastened and check the bump clearance. Reciprocating compressor owners and operators must adhere to "best torquing practices." They are crucial to the long-term reliability of components such as valve covers, for example. Good maintenance management should address the following issues pertaining to torque application in the field:

(a) *The friction factor*. There should be a clear understanding about the use of petroleum lubricants on threads before attempting bolt-up. Quantities and type of lubricant should be defined for consistency of results. There is an up to 35 percent reduction in bolt or stud stress levels obtained when a nut is torqued with a dry thread.

(b) *The operator*. Field technicians should have a good understanding of torquing sequence when bolts or studs to be torqued are arranged in a bolt circle.

(c) *Geometry*. Field technicians should be held to make gap measurements of a completed joint. Variations in circumferential gap measurements in 90° steps would be an indicator of uneven load distribution worthy of investigation.

(d) *Tool accuracy*. A testing and calibration schedule for torque wrenches must exist. If such a program is not implemented, the result of maintenance interventions involving the use of torque wrenches will become questionable. The more accurate the tool or stud tightening measuring system, the more accurately a bolt or stud can be preloaded.

(e) *Relaxation.* There is no doubt about a relaxation effect in any bolted assembly—it can be overcome by scheduled torque checks.

Follow up by checking the pipe load on flanges. Pipe supports often get bent or distorted and can put cylinders out of alignment. Finally, clean up the foundation and look for cracks. If any are visible, try to check the depth and provide a record for possible regrouting work during the next overhaul.

Major Overhaul Checklist for Reciprocating Compressors

1. Check clearance of main and connecting rod bearings.
2. Inspect crank pin, crosshead pin bushings, and crosshead bearings.
3. Inspect main bearing, shaft alignment, and crankshaft alignment.
4. Check alignment of crosshead and piston; readjust shoes, if necessary, using level on crosshead guide and cylinder base.
5. Check alignment of cylinder to crosshead and frame.
   (a) Mount dial indicator on cylinder housing and take reading on rod in horizontal and vertical direction through stroke.
      - If greater than 0.003"(76 μm) in vertical direction, correct by shimming crosshead shoes.
      - If greater than 0.003"(76 μm) in horizontal direction, check various fits and joints; correct as necessary.
   (b) Alternative to (a) in this list, use alignment wire or laser tool through centers of cylinder bore and crosshead guide bores.
6. Inspect all stud nuts for tightness (crankcase cover, distance piece, etc.).
7. Inspect, measure, and record piston and cylinder diameters.
8. Inspect, measure, and record piston rings, rider rings, and piston ring grooves.
9. Inspect valves.
10. Inspect oil coolers, water jackets, intercoolers.
11. Inspect lubrication devices, oil filters, sumps, and piping.

The most effective way to bring about seamless cooperation between maintenance and operations departments was conclusively demonstrated in the mid 1970s. This is when one of the co-authors reported to a plant manager. The manager was innovative and could not be fooled by people who merely generated noise. He told his maintenance department head and his top operations person that he, the plant manager, would call them within a year and tell them to trade their respective job functions. The operations manager would from that day on be the maintenance manager and the maintenance manager would be in charge of operations. The two cooperated from the first day to the last. They learned each other's craft and were ready for switching jobs within six months. The plant manager made the call about 8 months later and the job trading went flawlessly. The plant, the corporation, and everyone associated with these events was convinced it was a true winning strategy. Each of the two department manager had learned to become equally competent. They were able to demonstrate insight and success. They had learned the mandatory "3 Cs"—Cooperation, Communication, and Consideration.

What We Have Learned

There are overlapping responsibilities in compressor maintenance and operation. In general, data taking and spotting deviations from normal machine behavior is an operator responsibility. Problem correction and adjustments using other than simple tools is assigned to mechanical workforce members.

Shutdown procedures and work lists are developed by joint effort. All affected job functions contribute to this effort. The results are formalized in written procedures or checklists.

Bibliography

1. Bloch, Heinz P., and J.J. Hoefner, *Reciprocating Compressors, Operation and Maintenance*, Gulf Publishing Company, Houston, TX, 1996.
2. Leonard, S.M., *Reciprocating Compressor Upgrades for Improved Reliability and Reduced Maintenance*, Dresser-Rand Services Division, Painted Post, NY, 1997.
3. Miguez, J.P., *Problem of Compressor Over-Lubrication*, CompressorTechTwo, August–September 2007, 56–64.
4. Hatch, G., and D. Woollatt, Dresser-Rand Co. Predictive method improves compressor performance. *Pipeline Gas Tech.* January/February 2010, 40–45.
5. Bloch, Heinz P., *Practical Guide to Compressor Technology*, 2d ed, John Wiley and Sons, Inc., Hoboken, NJ, 2006.
6. Bloch, Heinz P., and F.K. Geitner, Practical Machinery Management for Process Plants: Volume 5, *Maximizing Machinery Uptime.*, 2006, www.books.elsevier.com.
7. Hickman, D., *Compressor Performance and Optimization—Staying within Rod Load and Pin Reversal Limits*, CompressorTechTwo, April 2007, 52–57.

CHAPTER 16

Surveillance, Monitoring, and Troubleshooting Reciprocating Compressors

Surveillance and monitoring of reciprocating compressors are terms used interchangeably. With few exceptions, the monitoring and/or surveillance tasks are assigned to the operators. They look for deviations from the agreed-upon operating behavior and take corrective action. *Corrective action* consists primarily of adjustments which require no tools other than perhaps a pair of pliers and a screwdriver. If remedial action involves the use of other tools, operating technicians are expected to interface with maintenance or reliability personnel. The work scope of maintenance personnel is to keep compressors at as-purchased performance. The work scope of reliability personnel is to view every maintenance event as an opportunity to upgrade. Reliability personnel must answer the question if upgrading—the removal of weak links in the component chain—is feasible. If they answer in the affirmative, it will be their responsibility to submit cost-justification data to management.

Surveillance

General

1. The key aspects of reciprocating compressor operation which require operator surveillance are:
   - Condition of compressor valves
   - Rod loading

213

- Rod packing vent emission
- Vibration and noise
- Lubrication—frame and cylinder
- Cylinder coolant
- Cylinder temperature

Valves

1. The common causes of valve failure are:
   - Dirt in the gas stream
   - Liquid in the gas
   - Insufficient cylinder lube oil
   - Gas pressure pulsations
   - Valve instability or "flutter"
   - Corrosive attack on valve material
   - Pieces of broken piston rings or other parts falling into the valve passages
   - Valve seat distorted by faulty installation

2. Indications of valve failure are:
   - High temperature at the valve cover
   - A reduction in capacity
   - Noise
   - An increase in discharge temperature

3. Change in the interstage pressure of a multistage compressor is an indication of valve failure:
   - If a valve on the lower stage fails, the interstage pressure decreases.
   - If a valve on the higher stage fails, the interstage pressure increases.
   - For compressors with unloading control devices, the interstage pressure for each stage of unloading should be recorded in the "as-new" condition, to provide a standard for monitoring.

4. Valves must be repaired as soon as they fail in order to avoid:
   - Scoring of the rod, piston, rings, and liner
   - Failure of the rod packing
   - Piston seizure or breakage

5. Valve repair is also necessary in order to restore:
   - Flow capacity
   - Operating efficiency
   - Normal cylinder-operating temperatures

6. The key to analysis of valve failures is an accurate record system which should include:
   - Port location of valve failure
   - Identity of the particular valve seat involved in failure

- Operating conditions preceding failures
- Nature and cause of failure, to the extent known

7. Operator monitoring of compressor piston rod loading is critical because:
   - Most major damage to reciprocating compressors results from excessive rod loading.
   - Rod failures can be hazardous to personnel and cause fires, as well as being costly to repair.
   - Rod load is not indicated directly by any single instrument sensing, and is not protected by alarm or trip features.
   - The rod load provides the simplest index of the loading level on other elements in the compressor power frame: head bolts, pistons, connecting rods, cross heads, bearings, etc.

8. To monitor rod loads, a "rod load limit" graph should be developed for each compression stage as shown earlier in Chapter 14 (Fig. 14.2). This graph should be posted in the operations control room.

Packing Vents

1. The emissions or drainage from the rod packing vent (or drain) is a key indication of cylinder's mechanical condition. A small amount of lubricant is emitted from the vent line.
   - Rate of gas emission varies with cylinder pressure level, packing design, and packing condition.
   - Since instrumentation is not provided to measure this gas rate, it must be judged subjectively.
   - Source of oil is rod packing lubricant and, to a lesser extent, cylinder wall lubricant.
   - Oil color and metal content reflect mechanical condition; dark oil color and/or bright metal particles in the oil indicate internal wearing.
   - On nonlube compressors, the gas emission rate provides an indication of rod packing condition.

Vibration and Noise

1. If reciprocating compressor cylinders vibrate, look for:
   - Loose supports under cylinders or piping
   - Loose studs or bolts on the machine
   - Deteriorated grout
   - Excessive rod loading
   - Speed or load above rating
   - Moisture entrainment in suction piping

2. Gas pulsations can also make reciprocating compressors vibrate, due to:
   - Pulsations created by operation in parallel with other machines
   - Speed higher or lower than optimum
   - Broken baffles in pulsation bottles
   - Faulty valve in one end of a cylinder

3. Some common sources of noise emanating from a malfunctioning reciprocating compressor include:
   - Expander rings of the piston, cutting the liner, emit a chattering knock
   - Tilting or rocking of the crosshead against its slider bearing emits a slapping noise
   - Liquid in the gas causes transient knocks of varying intensity
   - A piston striking one of the cylinder heads emits a heavy thumping noise

4. Knocking noises in compressor cylinder can also be caused by:
   - An unsecured cylinder liner sliding axially to impact against a head
   - Metallic debris within the cylinder
   - Excessive clearance in the crosshead bushings, connecting rod bearings, or crank shaft bearings
   - Loose valve assembly, loose packing, loose piston

5. To locate the source of noise:
   - Note that most noises appear to emanate from the crankcase, whatever their actual source may be.
   - Try to identify which cylinder or slider-crank assembly is producing the noise, if this can be safely done before stopping the machine.
   - Note severity of the noise, its character, and whether it is changing.
   - Allow no one to stand near or opposite the outboard end of an operating cylinder emitting unusual noises of any kind.

Lubrication

1. Crankcase lube oil must be monitored for:
   - Proper level—neither too low or too high, however, use numerical values in specific cases.
   - Temperature upstream and downstream of the oil cooler. Typical crankcase temperatures are $68°C \pm 3°$ ($155°F \pm 5°$); oil is supplied to the bearings at $60°C \pm 3°$ ($140°F \pm 5°$).
   - Oil pressure to the bearings: Falling pressure means pressure regulator problems, oil pressure safety valve problems, possibly bearing wear, or oil pump wear.
   - Typical oil filter pressure drop at which filter should be replaced or cleaned—1.4 bar (20 psi).

2. Monitoring of compressor and engine cylinder force feed lubricator system requires:
   - Confirming flow to all locations
   - Controlling oil flow rate at level determined to be optimum
   - Keeping the lubricant reservoir filled

3. Cylinder liner condition should be inspected for proper lubrication:
   - After break-in of a new or overhauled cylinder
   - After changing lubricant or oil injection rate
   - Each time a valve is removed for service

4. When inspecting cylinder liner conditions:
   - Liner surface should be uniformly bright.
   - Check 6 O'clock position for beginning of scuffing or scoring.
   - Lube film should be very light—almost imperceptible on a finger, but visible when wiped with cigarette paper or newspaper-type stock.
   - Confirm that each oil injection hole is open and functioning.

5. The effectiveness of cylinder wall lubricants can be drastically reduced by dilution with liquid in the gas stream, coming from:
   - Carryover from the suction knock-out drum
   - Condensation along the suction line
   - Condensation within the compressor cylinder suction valve chambers, due to excessive cylinder lubrication

Cylinder Cooling

1. When reciprocating compressors are provided with a circulating coolant for cylinder jacket cooling, the coolant flow should be carefully monitored:
   - Coolant supply temperature should be at least 6°C (10°F) warmer than the gas inlet temperature (preferably 8°C–11°C or 15°F–20°F) to avoid condensation within the cylinder.
   - Cooling water supply should be no more than 17°C (30°F) warmer than gas inlet temperature to avoid flow capacity loss.
   - The cooling water temperature rise across the cylinders should be at least 8°C (15°F), and preferably 11°C (20°F).
   - Cooling water outlet temperature should normally be in the 46°C to 63°C (115°F–145°F) range.

2. Typical clearances of a newly overhauled 180 mm (7 in) diameter cylinder are:
   - Piston to cylinder liner, 18 to 25 μm (.007–.010 in.)—overhaul needed at three to four times this level
   - Piston rod run-out (or "hop") in the running condition: 2.5 to 7.5 μm (.001–.003 in.)

Monitoring

1. Careful monitoring of the temperatures of reciprocating compressor cylinders is an effective way to:
   - Achieve maximum gas flow rate
   - Achieve high compressor efficiency
   - Reduce maintenance costs
   - Reduce unscheduled downtime
   - Permit scheduling maintenance work for convenient time periods

2. The two main concerns in monitoring reciprocating compressors cylinders are:
   - To avoid excessive discharge temperature
   - To find indications of deteriorating compressor valves and piston rings

3. Factors that affect cylinder discharge temperatures most are:
   - Inlet temperature
   - Compressor ratio
   - Specific heat ratio, k (c_p/c_v), of the gas
   - Mechanical condition of the valves and cylinders

4. Specific guidelines for maximum allowable compressor discharge temperature are:
   - See the manufacturer's operating and maintenance manual and the machine specifications.
   - Note distinction between maximum discharge temperature for which a new compressor should be designed (staging and intercooling design), versus the maximum discharge temperature at which a specific, existing compressor should be allowed to operate.

5. The areas of mechanical deterioration which cause increased discharge temperature are:
   - Piston rings and their mating cylinder liners—probably allowing blow-by
   - Compressor suction and discharge valves—probably allowing cyclic flow reversal in valve chamber area

Temperature Measurement

1. Methods of observing cylinder temperature include:
   - Control panel–mounted temperature indicators
   - Local thermometers
   - Portable surface or infrared temperature indicators
   - Hand touch—for identification of obvious valve failures

- Automatic temperature difference detection equipment—installed systems and portable instruments
- Electronic differential thermometer designed primarily for compressor valve monitoring

2. To identify cylinders which require maintenance, operators should compare:
   - Valve cover temperature against other covers of parallel valves on the same cylinder
   - Discharge temperature of parallel cylinders on the same compressor
   - Temperatures of discharge lateral pipes of parallel compressor units

3. Guidelines for evaluating differences between the discharge temperatures of cylinders operating in parallel:
   - 0.5°C (1°F) difference is a warning of possible deterioration.
   - 1.5°C (3°F) difference indicates a faulty condition in the cylinder with the higher temperature.

Efficiency Calculations

1. The discharge temperature is the key factor in the determination of the operating efficiency of a compressor cylinder. Two methods are recommended for calculating and tracking efficiency:
   - Polytropic efficiency, which allows for actual change in entropy of the gas during compression (a minor factor in most reciprocating compressors).
   - Isentropic (adiabatic) efficiency, which assumes no change in entropy during compression (a fairly accurate assumption for most reciprocating compressors).
   - Both methods are satisfactory for monitoring cylinder conditions. The choice will depend on the user's preference for the computations required. Applying both methods allows a double check.

2. Calculation of polytropic efficiency (η_p) as an indication of machinery health:

$$\eta_p = \frac{\frac{k-1}{k}}{\frac{n-1}{n}} \tag{16.1}$$

where η_p = polytropic efficiency
k = ratio of specific heats for the particular gas mixture (normally available in the design specification for the service, or from gas analysis)
n = polytropic exponent

$$\frac{n-1}{n} = \frac{\log\left(\frac{T_2}{T_1}\right)}{\log\left(\frac{P_2}{P_1}\right)} \qquad (16.2)$$

where T_2 = stage discharge temperature in degree Rankine (°R)
T_1 = stage suction temperature
P_2 = stage discharge pressure in psia (kPa(a))
P_1 = stage suction pressure in psia (kPa(a))

3. Predicted isentropic temperature rise with the k-value, compression ratio, and suction temperature known may be determined by using the chart, Fig. 16.1. It should also be noted

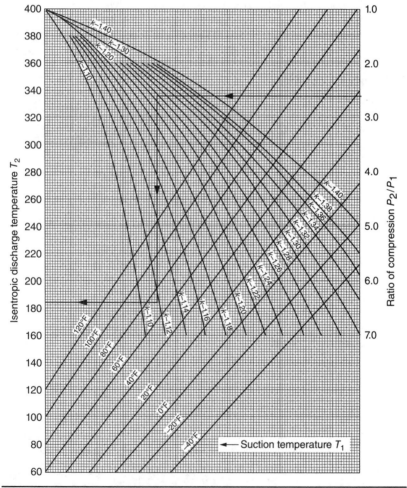

Figure 16.1   Predicting compressor discharge temperature.

that differences between efficiency values calculated by the polytropic and isentropic methods are small and unimportant when performing routine monitoring. However, it is important that a consistent method be used when comparing values. Helpful diagrams (Fig. 16.2) facilitate quick estimation of power draw as a function of suction and discharge pressures for certain defined reference conditions.

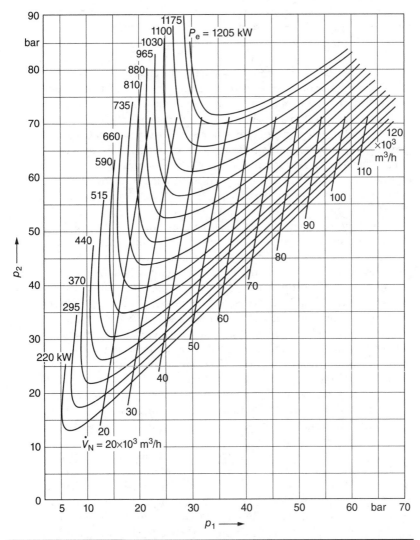

FIGURE 16.2 Panhandle diagram for a single-stage gas compressor at reference conditions PN = 1 bar; TN = 25°C. (*Source:* Borsig Gruppe Deutsche Babcock, Berlin. See also Karl-Heinz Küttner, Kolbenverdichter—Auslegung Betrieb Konstruktion, Springer Verlag, Berlin, Germany, 1992.)

Troubleshooting

Very few pieces of petrochemical process machinery lend themselves
to early symptom-cause identification as do reciprocating compressors.*
This becomes evident when considering how the neglect of subtle
performance changes can result in costly wrecks. For instance, high
discharge temperatures can result from the simple primary cause of
insufficient coolant supply. Not responding to this cause will lead to
an overheated cylinder and ultimately to such events as piston seizure, ring breakage, and piston cracking. Similarly, it would be a mistake not to respond to such audible symptom as "knocking in cylinder," where the most probable cause is inadequate piston-to-head
clearance. Inadequate clearance will ultimately lead to piston failure
or rod breakage with attendant damage to the crosshead.

When troubleshooting reciprocating compressors, the most important symptoms to watch for are unusual sounds and changes in
pressures, temperatures, and flow rates. Consequently, the primary
troubleshooting tools are our five senses, two pressure gauges, two
temperature indicators, and a flow meter. Generally, flow meters are
not available for each individual stage of compression, but considering
that what enters the front end will exit at the back end, flow measurement at one stage along the way is sufficient at installations where no
interstage inlets or knockouts are involved.

It is important to use a historical compressor log sheet to record
interstage pressures and temperatures by stage on multistage compressors. When interstage operating conditions vary from normal, it
indicates trouble with one of the stages. Generally, when the interstage
pressure suddenly drops, look at the lower pressure cylinder. If the
interstage pressure suddenly rises, look at the higher pressure cylinder.

Compressor temperatures and pressures are basic to design calculations and help determine compressor health. The difference
between the observed and calculated temperatures should be more
or less constant from day to day. The actual observed or calculated
temperature may vary; also, when suction temperatures increase, so
will discharge temperatures. If the compression ratio across a cylinder increases, its discharge temperature also increases. Comparing
calculated and actual discharge temperatures provides a means with
which to determine operating deviations.

Vibration analysis is usually applied to rotating equipment such
as centrifugal compressors. It can also be useful for reciprocating compressors. Vibration analysis can alert us to coupling misalignment,
even with relatively low-speed reciprocating compressors. Some
troubleshooters have used vibration measurements in the analysis of
piping fatigue failures and cracking or other fractures initiated by
inherent system resonance and pulsations. Some of these problems are

---

*Reference 2.

caused by first or second orders of running speed; either can excite resonant frequencies in pipe runs. Usually, if the vibration frequency is higher than twice the running speed of the machine, acoustic pulsations are suspected.

The following trouble shooting symptom-and-reaction listing is offered as a basic guide only. It lists the more common issues with their most probable remedies. It may be expanded to address specific installations, package configurations, and actual operating conditions.

Quick Troubleshooting Guide

Compressor Will Not Start

- Check power supply, switchgear, and the control panel.
- Verify proper oil pressure to ensure oil pressure shutdown has cleared.
- Ensure all starting permissive circuitry has been satisfied.
- Ensure nothing is blocking running gear or engine.

Low Oil Pressure

- Check all local pressure gauges for proper operation.
- Verify proper operation of oil pump.
- Inspect pump for wear, repair, or replace if needed.
- Check for proper frame oil level. Drain or fill as needed.
- Check lube oil filter differential pressure. Change elements as needed.
- Check setting of lube oil pump relief valve, increase if needed.
- Inspect frame lube oil strainer.

Frame Knocks

- Verify proper running gear clearances.
- Inspect running gear components.
- Check for proper oil pressure and temperature.
- Inspect cylinder, piston end clearance, piston nut, or crosshead nut.

Noise in Cylinder

- Check end clearance.
- Check piston and crosshead nuts.

- Inspect for broken or leaking valves.
- Inspect piston rings and rider bands.
- Verify valves are installed correctly.
- Verify unloader plugs are seating properly.
- Open discharge bottle drains to ensure no liquid in cylinder.

Excessive Packing Leakage

- Inspect packing rings for wear; replace as necessary.
- Check for proper lubrication and adjust accordingly.
- Inspect packing case for dirt or debris.
- Verify whether packing is properly assembled.
- Check packing vent line for obstruction.
- Inspect for damaged piston rod and repair or replace as necessary.
- Check packing clearances.
- Check piston rod run-out.

Packing Overheats

- Inspect packing for contamination.
- Inspect rod for damage.
- Inspect packing clearances, on the rod and in the packing case.

High Discharge Temperature

- Inspect suction and discharge valves for leakage.
- Inspect cylinder water jackets, clean if needed.
- Check suction start-up screen.

Low Suction Pressure

- Check inlet filter differential pressure, shift filters as needed.
- Inspect isolation valves for proper operation.

A more detailed analysis of the symptom-cause-failure chain may be in order and Table 16.1 can serve this purpose. Far more detailed information can be gleaned from a Bloch-Hoefner text covering this subject.*

---

*Refer to Ref. 1.

Symptoms

TABLE 16.1 Reciprocating Compressor Troubleshooting Chart

	Possible Causes	Noise Vibration – Knocking	Noise Vibration – Vibration	Pressure – Discharge Pressure Up	Pressure – Discharge Pressure Down	Pressure – Inter-Cooling Press. Up	Pressure – Inter-Cooling Press. Down	Discharge Temp. Up	Outlet Cooling Water Temp. Up	Temperature – Overheating Valves	Temperature – Overheating Cylinder	Temperature – Overheating Frame	Flow – Capacity Down	Int. Inspection Result – Carbon Deposits Abnormal	Int. Inspection Result – Piston Rings Cyl. Wear Up	Int. Inspection Result – Valve Wear/Breakage Up
Valves	L.P Valves Wear/Breakage				②	①	①	①	③	③	②		①	④		⑤
Valves	H.P Valves Wear/Breakage					①										
Valves	L.P Unloading System Defective		③	①	①		②	②	④	④	⑧		②	⑤		⑦
Valves	H.P Unloading System Defective					②										
Pistons/Cylinders	L.P Piston Rings Worn				④		⑤						⑦	⑨	⑥	
Pistons/Cylinders	H.P Piston Rings Worn					③										
Pistons/Cylinders	Piston Rod Nut Loose	④														
Pistons/Cylinders	Piston Loose	⑥														
Pistons/Cylinders	Head Clearance Too Small	②														

Note: The numbers indicate what to check first, or the probability ranking.

225

Symptoms

Possible Causes	Knocking	Vibration	Discharge Pressure Up	Discharge Pressure Down	Inter-Cooling Press. Up	Inter-Cooling Press. Down	Discharge Temp. Up	Outlet Cooling Water Temp. Up	Overheating Valves	Overheating Cylinder	Overheating Frame	Flow Capacity Down	Carbon Deposits Abnormal	Piston Rings Cyl. Wear Up	Valve Wear/Breakage Up
Frame															
Bearing Clearance Too High	⑤														
Flywheel or Pulley Loose	⑦	②													
Crosshead Clearance Too High	③														
Support/Cooling/Lubrication															
Cooling Water Qty. Too Low							④	①		④					
Cylinder Lubrication Inadequate	⑨	⑥					⑦			⑥				①	①
Frame Lubrication Inadequate	①										①				
Cylinder Lubrication Excessive							⑧			⑦			②		⑧
Lubricating Oil Incorrect Spec.	⑩												①	②	②
Foundation/Grouting Inadequate	⑧	④													
Piping Support Inadequate		①													

226

Table 16.1 Reciprocating Compressor Troubleshooting Chart* *(Continued)*

Piping/System														
Resonant Pulsations (Suction or Discharge)														⑨
Suction Filter Dirty/Defective			③						③		③		⑤	⑤
Suction Line Restricted			④						④		④			
System Leakage Excessive				③					⑤		⑤			
System Demand Exceeds Compressor Capacity				⑤										
Operating Conditions														
Discharge Pressure Too High	⑪	⑦	②			③	⑤	①	①	③	⑥			
Discharge Temperature Too High											⑦			
Intercooler Fouled				④		⑥	⑥				⑪			
Liquid Carry-Over											③		③	③
Dirt/Corrosion Products into Cyl.											④		④	④
Cylinder Cooling Jackets Fouled						⑤	②	⑤			⑩			
Running Unloaded Too Long								②						
Speed Incorrect		⑤		⑥		③		③	②		⑧		⑥	

Note: The numbers indicate what to check first, or the probability ranking.
*From Ref. 2.

227

Beta* analyzers allow us to directly recognize what is happening inside a compressor cylinder. This method of analysis, further explained in Ref. 2 of Chapter 5, can be extremely helpful in solving problems related to valve losses and piston-ring leakages. Such analytical equipment can be expensive, but there are companies supplying this service on either a contract or a one-time basis.†

Operators and maintainers of large fleets of reciprocating compressors in the oil and gas patch have been leading in establishing a record of successful automated performance data collection. They have implemented a system that reports process information from field stations, takes this data and turns it into usable information for both field operations and engineers. Operators at the compressor site enter operating data into customized checklists on handheld devices called portable data terminals (PDTs). When the operator is in range of a wireless internet connection, he or she can synchronize the data with servers over the web. A proprietary program instantaneously rationalizes the raw data, incorporates it into a user-accessible fleet-management database, and produces a detailed diagnostic report.

The report provides the operator with immediate feedback on the operating health of the compressor, complete with recommendations for both maintenance and production opportunities. The report helps the user to quickly see which compressors are operating inefficiently and which ones are at risk. The following beneficial results are claimed:

- Site-specific field inspection forms
- An electronic compliance calendar for scheduling
- Inspection events, and
- A corrective action–tracking system that allows operators and engineers along with outside specialists to assign tasks to resolve field deficiencies
- Delivery of an average benefit-to-cost ratio of 5:1‡

In spite of the above-described sophistication, we must in all cases remember that from a diagnostic point of view one-time readings are no substitute for recorded historical operating data. After all, the most decisive feature of a trouble symptom is its upside/downside *change*.

---

*Beta machinery analysis, www.betamachinery.com.
†See Ref. 3
‡See Ref. 4

References

1. Bloch, Heinz P., and J.J. Hoefner, *Reciprocating Compressors, Operation and Maintenance*, Gulf Publishing Company, Houston, TX, 1996.
2. Geitner, Fred, and Heinz P. Bloch, Practical machinery management for process plants, Volume 2, *Machinery Failure Analysis and Troubleshooting*, 4th ed, 2012, Elsevier, UK, www.books.elsevier.com.
3. Stachel, K., and C. Koers, Modern online monitoring systems for piston compressors. *Hydrocarbon Processing*, August 2003, 53–56.
4. Detechtion Technologies LLC. *Case Study: Talisman*, 2009, www.detechtion.com.

CHAPTER 17

Reciprocating Compressor Upgrading, Rebuilding, and Remanufacturing

Whether one's company must replace a reciprocating process compressor or an entire compressor fleet, it is always best to look at all available options before making such a substantial equipment investment. Compressor owner-operators are sometimes surprised how cost-effective it can be to buy a remanufactured compressor. Naturally, there are upsides and downsides to this approach and we should be familiar with all reasonable options.

There are minor and major distinctions between the various definitions. Many definitions overlap, but still deserve to be listed:

- *Inspection*: An activity aimed at determining which compressor parts are in need of repair
- *Overhaul*: Reconditioning and repairing of what needs to be repaired, bringing into compliance with perceived original condition
- *Repair*: Replacing parts that have deteriorated or are found defective
- *Rebuilding*: Dismantling, disassembly, and reassembly involving a compressor and its mounting location
- *Refurbishment*: Same as rebuilding, but including auxiliary support system upgrades

- *Upgrading*: Strengthening parts that fail more often than other parts; systematically removing the weak links in a component chain so as to extend availability and reliability

- *Conversion*: Implementing changes to accommodate major differences in service or application

- *Rerating*: Modifications with the goal of reducing rated throughput for the purpose of gaining mechanical efficiency, or increasing throughput capability to gain production capability

- *Revamping*: Modifying for the purpose of adapting to new process conditions

- *Reapplication*: Same as revamping, but often involving compressors that had been out of service for some time (mothballed)

- *Remanufacture*: Total strip-down into every component, nut, and bolt, full and uncompromising measurement and/or testing of every part

Here is more detail on some of the activities mentioned.

Inspection, Overhaul, and Repair

Besides operating, the normally expected activities around reciprocating compressors are inspection, overhaul, and repair (IO&R); the other functions listed above are extraordinary measures as the machine moves through its life cycle. Historically, IO&R was conducted in refineries and chemical plants by in-house "captive" maintenance forces supported by a central machine shop. These programs were self-directed and were carried out with the occasional assistance of OEM field service personnel.

As time went on, captive maintenance forces were reduced and supplemented by outside contractors. In a typical setting, an experienced company technician would lead a group of contract mechanics in carrying out the required IO&R. Starting in the late 1990s and as this text went to press (2012), many owner-operators of reciprocating compressors have resorted to completely contractor-conducted IO&R programs. Depending on the sophistication of owners or contractors, either periodic or predictive maintenance strategies are followed to determine IO&R intervals. Due to cost considerations, upgrading and concurrent rebuilding of compressor parts as opposed to purchasing expensive OEM spare parts became a normal practice. The compressor world was therefore not surprised more than a decade ago when a reputable major upstream reciprocating compressor manufacturer[*] announced it had discontinued its new unit business and would forthwith only sell remanufactured machines.

---

[*]Cooper-Bessemer Corporation

Rebuilding

The rebuild process takes existing components or entire compressor units and returns them to the original factory specification as part of a preventive maintenance program. This would reduce the risk of potential compressor failures to occur during peak operating business cycles; the practice is a natural outflow of the overhaul and repair routine. It was a normal procedure when compressor fleet owners had access to in-house repair shops. The cost to rebuild existing components of reciprocating compressors and even rebuilding entire machines is certainly lower than the cost of replacing them with new equipment. Additionally, the owner-operator controls the downtime schedule and selects its timing to have the least impact on business. An example is the rebuilding and upgrading of piston rods in the maintenance environment as demonstrated by the topics expounded in a repair specification, Fig. 17.1. An actual specification would comprise many pages and would become the discussion basis for maintenance, repair, and rebuilding activities.

Compressor Piston Rod Repair and Coating

1.0 SCOPE

This specification applies to the inspection, repair, and rebuild of reciprocating compressor piston rods. The portion of the rod that operates in the packing shall be restored with a high velocity—applied tungsten carbide overlay.

2.0 Rod description and special instructions

3.0 Inspection

4.0 Precoating preparation

5.0 Rod coating

6.0 Finish

7.0 Records, preservation, and shipment

8.0 Inspection

FIGURE 17.1 Topics covered in maintenance, repair, and rebuild specification.

Upgrading, Conversion, Rerate, Revamp, or Reapplication Decisions Resulting in Rebuilding, Refurbishment, and Remanufacturing of Existing Reciprocating Compressors

Upgrading, converting, rerating, revamping, or reapplying existing equipment can help meet new or changing compression requirements with existing equipment. This activity then provides a cost-effective alternative to the purchase of new equipment and projects can generally be completed in a shorter time period. Additionally, this approach can extend equipment life, increase reliability, and improve availability by incorporating the latest material and design technology. Finally, it reduces compressor cylinder fugitive emissions to meet current and future environmental requirements.

Competent corporations* recognize the interaction of hardware and software. They typically specialize in custom compressor cylinders and unloaders as well as interactive performance software. One leading supplier is also known for providing performance-enhanced compressor components such as pistons, rods, liners, and custom valves as well as engineering services that include thermodynamic performance reviews, acoustic studies, and more.

A case in point involved the conversion of valve-in-head cylinder designs—see Fig. 17.2—to a more reliable and cost-effective configuration. The three-piece cylinder body was a common design traditionally supplied by many compressor OEMs decades ago. However, in this traditional design it took time to access the pistons, and their large size meant dealing with heavy masses. The old cylinders incorporated a large quantity of threaded fasteners in relatively inconvenient locations. Finally, gasket material discontinuations had made it increasingly difficult to obtain reliable and safe gaskets.

The revamp case involved a U.S. pipeline company in 2007 with the Cooper-Bessemer GMVA-10 integral engine compressor shown in Fig. 17.3. The machine was used in natural gas storage/withdrawal service with a maximum working pressure (MWP) of 1200 psig (82.7 bar).

The upgrade solution centered on providing cast ASTM A395 ductile iron cylinders of the type shown in Figs. 17.4 and 17.5. More specifically, it also involved

- A total of (16) cylinder replacements on (4) compressors
- A valve-in-barrel design which eliminated large gaskets
- A full-fledged bolt-in replacement approach
- Using and refurbishing existing valves and packing
- Rerating and refurbishing the existing clearance pockets
- Employing a water-cooled, jacketed cylinder design with a bore diameter of 12.0 in (304.8 mm), stroke: 14.0 in (355.8), testing compliance for an MWP of 1200 psig (82.7 bar)

---

*www.aciservicesinc.com/; by permission

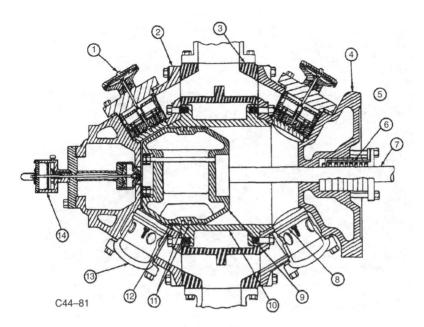

C44–81

1. Pneumatic finger type
   Suction valve unloader
2. Head end head
3. Compressor body
4. Crank end head

5. Suction valve
6. Rod packing
7. Compressor rod
8. Discharge valve
9. Piston

10. Cylinder liner
11. Piston rings
12. Rider rings
13. Valve cap
14. Head end unloader

Typical Valve in Head Compressor Cylinder

FIGURE 17.2 Three-piece cylinder. (*Source:* Cooper Bessemer.)

FIGURE 17.3 Three-piece cylinders before revamping. (*Source:* ACI Services, Inc., Cambridge, OH.)

FIGURE 17.4   Valve-in-barrel cylinders on revamped compressor. (*Source:* ACI Services, Inc., Cambridge, OH.)

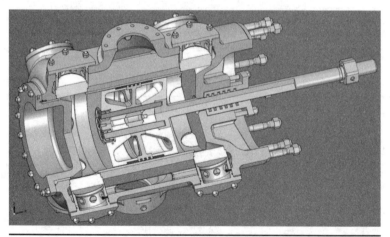

FIGURE 17.5   Model of a valve-in-barrel cylinder assembly. (*Source:* ACI Services, Inc., Cambridge, OH.)

In essence, the new and improved design resulted in higher reliability and safety by providing gasket elimination which reduces possible gas and coolant leaks. It can be reasoned that leakage avoidance increases plant safety. Fewer and more accessible fasteners reduce maintenance effort and cost of both labor and downtime. Considerable savings accrued by reusing most existing cylinder components. The revamp or upgrade approach preserves the value of an existing spare parts inventory.

It demonstrably minimized replacement cost by reusing existing pulsation bottles and mounting locations. Finally, it minimized the required installation downtime.

Remanufacturing is defined as an industrial process in which worn-out products are restored to like-new condition. Through a series of industrial processes in a factory environment, a discarded product is completely disassembled. Useable parts are cleaned, refurbished, and put into inventory. Then the product is reassembled from the old parts—and where necessary, new parts—to produce a unit fully equivalent and sometimes superior in performance and expected lifetime to the original new product.* This would be in contrast to *rebuilt or repaired and repainted like most others.*

The following benefits of remanufacturing are claimed:

- The whole compression unit undergoes a comprehensive process to ensure reliability and performance
- Restored to like-new condition
- Meet or exceed original equipment manufacturer (OEM) new equipment specifications
- Same warranty as new
- Receive full mechanical and performance testing
- Shorter lead times
- Reduced downtime
- Cost savings
- Lower energy costs
- Latest technological upgrades
- Increased productivity

Remanufacturing of reciprocating compressors is often carried out by the OEM, but there are numerous independent service shops taking on the job.[†] Some of these non-OEM companies have been supplying their customers with new and remanufactured reciprocating compressors for almost 30 years. Especially in the high-speed reciprocating compressor after-market we find large inventories of used and rebuilt major compressor components such as frames, crankshafts, cylinders, and pistons.

Of course, the OEMs are not staying idle. One OEM[‡] states that they often work with equipment or parts the user or owner has already on hand, even if they did not build them. This strategy allows

---

*Lund, Robert T. Remanufacturing. *Tech Rev*. Feb–Mar 1984, 87, 2, 19–23, 28–29.
†www.sinorengine.com/httpdocs/index.php; www.lonestarcompressor.com; www.revak.com/.
‡www.dresser-rand.com/service/engineeredsolutions/revamps/recip.php.

owner-operators to place idle assets back into production and thus reduce the cost of doing business. Should the owner not have the appropriate equipment, the OEM claims to be able to locate it for him. If the equipment is out there—anywhere in the world—the OEM offers to find it fast.

Once a suitable unit is located, the OEM completely overhauls it to bring it up to the specifications required by the purchaser's application and delivers it to the owner with service manuals and a parts and labor warranty. The OEM offers to keep the purchaser informed during each stage of the remanufacturing process. The OEM also is usually able to provide field installation and start-up services for a remanufactured unit.

An example: When an air separation client asked an OEM to reapply a reciprocating compressor from nitrogen service to hydrogen service on the Gulf Coast, there was no time to waste: the client had an opportunity window for their product stream and the equipment needed to be ready at the appropriate stage of the new site construction.

The plan was to remanufacture the client's compressor frame and install new cylinders. Compared to purchasing a new unit, this would reduce costs, shorten the delivery schedule, and reutilize an existing production asset.

While the frame was being remanufactured at one of the OEM's service centers, the cylinders were being engineered and built at another location. The cylinders were shipped to the service center and mounted on the frame. All system tests and inspections were completed before the equipment was shipped to assure that all components fitted together correctly, all tolerances were met, and all systems were functioning properly. This simplified field installation and minimized the risk of costly delays—especially important since the construction delays had cut the installation time in half.

As a minimum, the purchaser of a remanufactured compressor should expect the following essential steps:

- Disassembly
- Inspection and tear-down report
- Parts to be replaced
- Assembly
- Leakage test
- Test run
- Painting

The most important step is an appropriate disassembly and inventory taking process followed by the above inspection and tear-down report as listed in Table 17.1.

Compressor Components	Remove	Open	Disassemble	Disconnect	Clean	Flush	Inspect	Record Condition	Label	List	Sketch	Test & Calibrate	NDT* & Surface Condition	Assess	Photograph	Repair	Rebuild	Replace	Checklist	Reassemble	Record Clearances	Record Run-out
Unit Received from Field															•				•			
Main Frame	•	•	•				•								•				•			
Main Bearings		•	•		•		•	•						•	•				•			
RTDs Main Bearings			•	•			•[1]	•					•	•					•			
Crankshaft							•	•					•	•[2]	•	•			•			•
Journals							•	•			•			•	•							
Direct Drive Lube Oil Pump Tubing/Piping				•			•	•	•	•												
Lube Lines				•				•			•											
Connecting Rods			•				•[1]	•				•[3]		•	•							
Crosshead Shoes & Slides							•	•					•[4]	•	•				•	•		
Crosshead bearing housing			•				•	•					•[2]	•					•	•		
Pistons	•		•				•	•				•	•	•	•							
Piston Rod Fastener (Nut)	•		•				•	•						•								
Compression Rings	•		•				•	•	•					•[5]								

TABLE 17.1 Tasks After Receiving Compressor Packaged Unit to Be Rebuilt or Remanufactured

Compressor Components	Remove	Open	Disassemble	Disconnect	Clean	Flush	Inspect	Record Condition	Label	List	Sketch	Test & Calibrate	NDT* & Surface Condition	Assess	Photograph	Repair	Rebuild	Replace	Checklist	Reassemble	Record Clearances	Record Run-out
Rider Bands	•		•				•	•	•					•[3]								
Piston Rods	•		•				•	•		•	•			•	•							•
Wipers & Packing Cases	•		•				•	•					•[6]	•								
Cylinder Bores/Liners							•	•						•								•
Cylinder Coolant Jackets					•	•	•	•				•[7]		•								
Cylinder gas channels							•	•				•[8]		•								
Inspection Side Doors	•	•	•																			
Distance Pieces Internals			•		•		•	•					•	•	•							
Distance Piece Registers							•	•					•	•								
Tubing—Distance Pieces			•	•			•	•	•	•	•											
Piping—Distance Pieces			•	•			•	•	•	•	•											
Valve Covers	•		•				•	•														
Unloaders	•						•	•														
Valves	•			•	•		•			•												
Fasteners/Grade									•									•				
Ancillary Equipment																						
Lube Oil Pumps							•					•		•								

TABLE 17.1 Tasks After Receiving Compressor Packaged Unit to Be Rebuilt or Remanufactured (Continued)

Component												
Coolant Pumps	●					●			●		●	●
Filters	●		●	●	●	●			●	●		
Filter Elements	●		●								●	
Flow Control Valves			●	●	●	●			●			
Pressure Control Valves			●	●	●	●			●			
Vibration Transducers	●		●		●				●			
Vibration Transmitters	●		●		●				●			
P/T/L Transmitters	●								●			
Instrumentation & Controls	●		●	●	●	●	●		●			
Valve Unloaders	●	●	●	●	●	●			●	●	●	●
Lube & Coolant Lines	●		●	●			●		●			
Electric Conduit	●		●	●					●			
Suction Piping	●		●	●	●	●			●			●
Discharge Piping	●		●	●	●	●			●			●
Knockout Drums	●		●	●	●	●						●

*NDT = nondestructive testing PMI (positive metal identification), crack & flaw detection, ultra sound, eddy current, magnetic particle detection, dye penetrant, straightness, surface conditions).

[1]Check for clear lube oil porting to main and connecting rod bearings.

[2]Perform web deflection.

[3]Test for flatness and clear lube oil porting.

[4]Check for clear lubrication ports.

[5]Assess design and material of construction as to suitability for operating conditions.

[6]Examination of all packing cases and wiper/scraper glands.

[7]Hydrotest.

[8]Watch for corrosion and/or restrictions.

It is important that, upon receiving the unit, the loads of the machine are evenly distributed. The compressor should be blocked or shimmed to obtain as close as possible level condition. Caution should be exercised when lifting and moving the unit in order to avoid stress loads in areas that are not designed to carry the load. Individual check lists should be prepared for all measurements and procedures used in the rebuild and remanufacturing process.

Modern coupling alignment tools are a must for a typical compressor remanufacturing facility. It would offer laser alignment tasks for any type of coupled equipment.

Modern bore alignment equipment is another requirement. Along with coupling alignment, laser tools are capable of measuring bores in internal combustion engines, compressors, compressor cylinders, and other machinery from 5 to 30 in (127–762 mm) diameter. A bore alignment tool is used to measure crankshaft bores. Once measurements are taken, one laser transmitter is set outside the last bore. A laser receiver is then rotated in each of the bores. The computer calculates the straightness and location of each bore within 0.0001 in (2.5 μm). It also measures alignment of multiple bores and are in-line with one another and in-line with the rotating centerline of the shaft. The measurements should be traceable to ISO 9000/MIMOSA standards. Complete reports can be printed instantly on location.

Contact with Service Shops*

The person or persons responsible and accountable for reciprocating maintenance should establish contact with service shops. This is best done by visiting them and judging their facilities, also their "track record" and the experience of their personnel. This review could lead to a numerical rating on a scale of 1 to 10 help with the final decision.

It goes without saying that quotations for new equipment prices should be obtained, so that the practicality of a repair, rebuild, or remanufacturing order can be ascertained. For instance, as a rule of thumb it would not be advisable to have an electric motor rewound if costs exceeded 70 percent of a new equivalent replacement, or if higher efficiency replacement motors are available. Also, if time is available, the purchase of surplus equipment may be considered. It would be advisable to maintain a subscription to at least one used or surplus equipment directory for that purpose. Also, there are many listings to be found on the internet.

The machinery maintenance person, during a facility visit, should gather information as to what procedures the shop uses to comply with his site or plant specifications. Obviously, a final sourcing decision should be made only after analyzing all available data and after the visit. The analysis can be made in form of a spreadsheet, using a

---

*See also Ref. 1.

marking pen to highlight pertinent facts and color-coding prices by relative position. In essence, this rigorous procedure is similar to a formal bid-evaluation process and would rank the bidders by shop capacity, experience, reputation, recent performance, order backlog, or even labor union contract expiration date and the like.

OEM versus Non-OEM Compressor Rebuilding, Refurbishment, Upgrading, Conversion, Rerating, Revamping, Reapplication, and Remanufacture

Equipment users are inundated with reams of technical information concerning machinery in the purchasing phase. Yet, seldom do operators/users get an opportunity to ask some very basic questions that deserve to be answered to run their business, and even less information is available on rebuilding, refurbishment, upgrading, conversion, rerating, revamping, reapplication, and remanufacture.

When it comes to remanufacturing of reciprocating compressor, one should build up a working knowledge of what to expect before one enters into the contingent transaction. Ask the potential supplier the following questions:

- Are the compressors remanufactured, or are they simply "rebuilt" or "repaired and repainted"? As we saw, completely remanufactured compressors are torn down to the smallest components and actually manufactured, so they should receive a similar guarantee to a new one. Ask about the manufacturer's engineering quality standards and what type of technology is used for the remanufacturing process.

- What is included with the remanufactured compressors? Is the quoted price for the actual unit only, or does it include the driver, instrumentation, control panels, heat exchangers, and other necessary equipment? If one ends up having to pay the manufacturer extra for all of the auxiliary equipment, one may discover that one actually has to spend more for that company's remanufactured compressors.

- How quickly can the remanufactured compressors be delivered? You should expect the delivery time for these remanufactured products to be significantly less than new compressors. Manufactures who state that it will take longer to fill these types of orders should be closely examined.

- Will the API 618 standard be followed and if not, what would be the exceptions to that standard?

- What other industry standards will apply instead?

A formal and active quality system is mandatory for any repair, rebuilding, refurbishment, upgrading, conversion, rerating, revamping, and remanufacturing facility. This implies that an all-encompassing system to control all the activities of an organization is in place to ensure that what is shipped is exactly what was ordered. Included in such a system are the following functions as a minimum:

- Formal organization and control
- Control of documents
- Calibration of instruments
- Calibration of tools (e.g., torque wrenches)
- Training and qualification of personnel
- Product identification and traceability
- Corrective action

What We Have Learned

In the final analysis, the compressor owner, user, or purchaser of outside inspection, overhaul, repair, rebuild, refurbishment, upgrade, conversion, rerate, revamp, reapplication, and remanufacturing services should apply forethought. The same principles of prudent decision-making apply as are exercised on the acquisition of new compression equipment.

In a best-of-class facility, the various aspects of compressor performance are intertwined and viewed as the responsibility of job functions that include process/operations, mechanical/maintenance, and project/technical.

Continuity of effort and quality of work require development and adherence to written standards, checklists, and work procedures. A facility has to groom and develop personnel who are the "owners" of a compressor. These employees will have authority and accountability. They, and others, will benefit from reading, rereading, and carefully considering this text.

Reference

1. Bloch, Heinz P., and Fred K. Geitner, Practical machinery management for process plants, Volume 3, 3d ed, *Machinery Component Maintenance and Repair*, Gulf Professional Publishing, an Imprint of Elsevier, Amsterdam, www.books .elwsevier.com, 2005.

CHAPTER 18

Training Competent Compressor Engineers

To achieve real proficiency and high productivity, professional and craft employees require structured training. Plants and facilities that allow training to be haphazard and unstructured have very often become unprofitable. As their deficiencies were then laid bare, entire training departments have been dissolved and various elements of self-training or on-the-job training were chosen instead. But the challenges often remained. Sound implementation of meaningful and effective technical training requires forethought and consistency. Loosely defined and sporadically executed self-teaching routines are of little value; structure, repeatability, and value-adding are needed for good training.

Role Statement and "Phase One" of In-House Training

The beginning of structured training should be a well-focused, written role statement which explains to both manager and managed their respective perceptions of the technical employee's role. The compressor professional or reliability technician may be encouraged to develop his or her own role statement and then ask for the supervisor's or manager's input or guidance. The last thing either party would want is a misunderstanding as to the perceived role.

Consider the mandate under which we operate. Are we parts changers or innovators? Are we expected to be fixers or improvers? A person who is expected to react to problems or anticipate problems? The role statement must, at least, allude to a training plan. The technical person and his or her supervisor should discuss both role statement and training plan initially and, of course, during scheduled future performance reviews.

The detailed training plan may well be a separate document. A good one will give firm guidance and yet leave lots of room for individual initiative. Its aim will be the achievement of proficiency in a technical skill or craft. As an adjunct, the employee is asked to keep a small computer file or notebook/folder with very brief notes on work performed and documenting the timelines of his or her achievements.

Nine out of ten times the resulting record of one's work product will be of great value. Pity the professional who will say he or she was too busy working and didn't find 2 minutes at the end of the week to write down what was accomplished.

Back to the training plan and, by way of example, a description of how technical training for a young engineer could be structured.

The Structure of Training

Let's say your facility employs four maintenance/machinery engineers or reliability technicians. You could get them to engage in worthwhile self-training by obtaining subscriptions—many of them electronic and free of charge—to such periodicals as

- Chemical Engineering
- Chemical Processing
- Compressor Tech2
- Diesel Progress
- Hydrocarbon Processing
- Machine Design
- Machinery Lubrication
- Plant Services
- Power
- Sound and Vibration
- Turbomachinery International
- UPTIME Magazine
- World Pumps

and many others that were available in 2012 and might still be around in the coming decades. Each technical employee is considered a "professional-in-training"—a PiT. His or her name would be at the top of a routing or PiT distribution list for two or three different periodicals. The designated professional would be required to screen the content of the monthly periodical(s) for relevant material, which takes about 5 minutes per month, in most instances.

We are careful to emphasize that an employee would not have to read the various articles, but would be expected to recognize from headings or abstracts the present or possible future usefulness of the write-up. Electronic copies of relevant write-ups would have to be made or simply passed on to the other "professionals-in-training" on the "PiT" distribution form. Copies of appropriate articles would be filed so as to be accessible to all; the articles or write-ups would be under appropriate headings. The file system—electronic or paper-based—might follow a simple, but logical numbering system to enable easy retrieval. This retrieval typically involves a straightforward, extensively cross-referenced, PC-based software program.

The value of collecting interesting articles can go well beyond what is commonly perceived or anticipated. Think of finding an equipment-related article written years ago on a certain subject, or by a particular author. Perhaps you have never heard of the author, but once you find the article in your old files, you will know if the writer is a man or a woman. He or she certainly is a communicator, a researcher, a person who spent a certain effort to set himself or herself apart from the rest of the profession. You locate this professional via the Internet and explain that you have—right in front of you—his or her old article on, say, rules-of-thumb for butadiene compression. You explain that you're a young engineer struggling with the concept of limiting Mach numbers on multicomponent heavy syngas and were wondering if this solid professional could please aim you in the right direction or spend 2 minutes explaining a certain issue.

Chances are that this author will respond very favorably and will be of immediate assistance. That's networking at its best; it represents connecting with the talent of your most competent predecessors or peers. And think of it: All it took was saving an old article!

"Phase Two" Training: Digging Up the Facts

The second phase of training might be called the "dig-up-the-facts" phase. Decades ago, one large multinational petrochemical company called it "the shirt sleeve seminar phase" because the employee would have to roll up his or her sleeves and teach others about plant issues and things everyone needs to know for improved proficiency.

Each "PiT" would be asked to present such periodically scheduled briefings or information-sharing sessions to mechanical workforce personnel. These workforce members were ones assigned to shop or field (e.g., millwright) tasks. The PiT would make his or her 7 to 10 minute presentation at the conclusion of the ubiquitous plant safety meetings. Decades ago, these highly effective presentations were also described as technology briefings or information sharing sessions. They would deal with such topics as "How to Install Rolling Element Bearings," "Proper Startup Procedures for Compressors and Drivers," "Why Four Different Types of Couplings are Used at our Plant," "When to Use Dry Gas Seals Instead of Oil Film Seals," "Things that go Wrong on Centrifugal Compressors," and so forth.

At one best-of-class facility, the abbreviated narrative of the PiT's presentation was captured on a single, two-sided, standard size sheet. Illustrations were pictured on the same sheet. These two-pagers were then laminated in plastic, three-hole-punched, and handed out to each attendee at the end of the shirt sleeve seminar.

An image of one of this plant's compressor couplings (Fig. 18.1) illustrates the point. It shows a diaphragm coupling fitted with a stretch-monitoring decal used at the PiT's facility. Incorporating narrative and sketches into a shirt sleeve seminar presentation educated the presenter, taught his or her audience, and served as a cornerstone for technology transfer to the next generation of engineers at this plant.

There are literally hundreds of worthwhile topics to research and discuss and disseminate. The "digging-up-the-facts" process would compel the presenter to do some homework instead of guesswork, communicating with vendors, other company staff and manufacturers instead of reinventing the wheel. The presenter might perhaps rediscover one or more of the many good technical textbooks which are generally available at a fraction of the cost of making even a single mistake. In the process of doing some diligent reading, the reader-researcher would inevitably educate himself or herself. He or she would, at the same time, contribute significantly to the development of team spirit and the enhancement of mutual respect and cooperation among the many job functions in the plant.

From here, the phased approach to training could move to in-house courses taught by competent presenters with both analytical and practical knowledge in compressor maintenance and reliability improvement procedures, and then progress to well-defined, known-to-be-relevant outside seminars or symposia.

And, by the way, if you think training is expensive, try calculating your costs *without* proper training. Without this training it will be impossible to retain talent and retain a loyal and motivated workforce.

Training Plans Beyond Phases I and II

Not to belabor the point, but reading is a hugely important part of training. That said, professional employees are encouraged to purchase texts that spell out solutions to problems. If a $100 text describes how to solve a $100,000 problem, or if investing 2 hours on reading will avoid spending 200 man-hours on repairs, the benefits should be abundantly clear.

Books purchased with company funds are company property and must be made available to other professionals. Funding for the books comes from the employer; investing personal time in reading represents the employee's share.

Attending on-site or off-site training courses is Phase III of the structured plan. Successful course attendance must be quantified

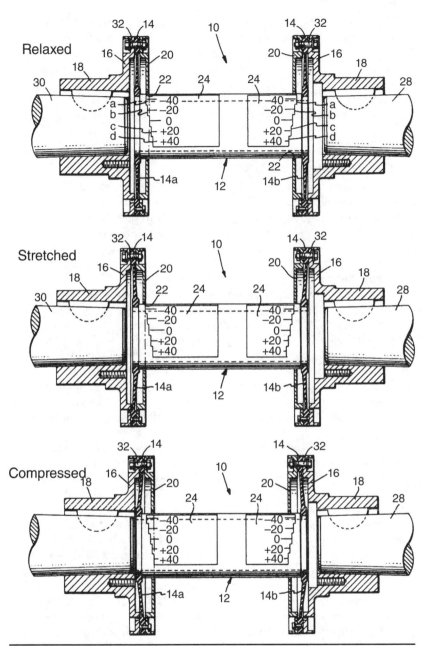

FIGURE 18.1 Cross-section of a diaphragm coupling fitted with stretch observation decal. Together with appropriate narrative and laminated in plastic, it typifies handouts from shirt sleeve seminar used for training and knowledge transfer (Ref. 1).

by achieving certification. After returning from such training, the employee highlights things learned in 250 words or less and shares the experience with the manager and other colleagues or interested parties. Consider this branching out the beginning of a life-long knowledge sharing and networking experience. For professionals with writing skills, this branching out then progresses to writing an article for one of the trade journals accessed in Phase I.

Putting the Training to Practice

There is a fitting conclusion to this topic; it is about engineers taking risks and communicating these risks to management. This text contains much information on failure analysis and troubleshooting. Try not to explain only the problem, but proceed to outlining the solution to the problem. After the cause is determined, the results need to be communicated to management on what the problem is and what has to be done to fix it.

As engineers we like to limit our risks. A highly experienced colleague, general aviation pilot and mechanical engineer observed how this has served him well over the years.[2] He does not recall doing things that were too risky, and he always had a couple of alternative plans in case something went wrong. His requests for a design modification were always supported with adequate calculations. He knew well that when engineers have performed a reasonable analysis, their arguments carry far more weight than the pleas of employees who are speculating on causes without bringing with them supporting data.[3]

There can be problems with the speculative approach; he learned these lessons early in his career and had to make adjustments. Again, the first observation is about risk: There is always risk involved in an engineering decision, and engineers cannot progress far in their careers if they are unwilling to take at least some risk.

Consider a large steam turbine driver vibrating slightly above normal levels with blade fouling thought to be the problem. Management wants to know if the compressor train can run an additional week before a planned outage with that level of vibration. Your career will not be enhanced if you say it has to be shut down immediately, with no supporting data. Likewise, this is not the time to try your first attempt of on-line washing of a steam turbine while it is in operation. It's risky business if you have neither experience nor operating guidelines for this procedure. However, this would be a good time to monitor the vibration level, talk with the manufacturer and others who operate similar machines, and determine the risk in just monitoring the vibration levels. Defining at what level the turbine will have to be shut down will still require some risk, but now others have given their valuable input. Obviously there is much more involved, but this illustrates the need for taking a certain amount of risk. You can expect to make errors, but these errors should not occur very early in your career and should also not occur without letup, one after the other. By the time the next error is made, you should have established a record of sound reasoning and mature judgment. Your occasional misjudgments would be forgiven; your sound contributions would outweigh your infrequent misses.

Correctly communicating to management what has occurred and what needs to be done is so important to an engineer. It would be wonderful if engineers had the verbal ability of attorneys in presenting data to management. Attorneys' jobs and training include making juries feel comfortable with what they are telling them, the decision that has to be made with the evidence, and data they have. Regrettably, many of us don't have this kind of training. Fortunately, it can be learned by experience and watching other successful engineers. Your company's senior technical personnel didn't get where they are by displaying a lack of communication skills or lots of poor judgment.

There are three things this source has found important when discussing work with senior management. The first two items are self-explanatory, but the last will require an example.

1. Management does not like to hear bad news, so present positive plans.

2. Management does not want to hear a wish list of solutions, so present only your best and most cost-effective choice.

3. Management is not impressed with complex analyses or unfamiliar technical terms; therefore, the engineer must simplify explanations of the cause, of the solution, and of the implementation so it can be understood and acted on.

The managers to whom you are presenting data and recommendations may not be familiar with mechanical things or rotating machinery engineering; their expertise may be in completely different fields. It is useful to adjust what you are presenting to suit your audience.

Suppose you are discussing the resonance of a structure and its failure. Now, as engineers we know resonance can be a highly damaging vibration caused by exciting a structure's natural frequency. Resonance could be clearly demonstrated by bringing a tuning fork to the meeting, striking it, and then showing the vibration behavior and sound emitted by the tuning fork. With no continued exciting input (meaning striking it again), the vibration dies out. However, with continued input and without a surrounding material damping the vibration, the tuning fork would fail in fatigue. The fatigue failure action could be demonstrated by bending a paper clip back and forth to show and explain what also happens in an industrial component.

At any point you can stop bending, but some of the paper clip's life has been used up.

When reviewing a technical presentation, the expert mentioned earlier always made a list of all the questions he thought might come up during the meeting and he researched them thoroughly. This allowed the best possible answers for the audiences with whom he met. His efforts were valued by those who recognized value. He managed to teach those that were teachable. Whenever he ran into the few that are either unreasonable or simply not teachable, he was able to shrug them off. You, the reader, should feel free to copy his example in assessing risks and making your presentations to management.

What We Have Learned

- Training is a shared responsibility between employer and employee.

- The progress of training must be monitored by the employee's supervisor and a management sponsor.

- Training progresses in phases and is an indispensable ingredient to professional growth.

- Shirt sleeve seminars are a zero-cost means of training presenters, audiences, and future generations of employees.

- Presentations delineating solutions enhance careers; presentations that only describe problems are not helpful.

References

1. Bloch, Heinz P., United States Patent 4,102,052; *Deflection Indicator for Couplings*, July 25, 1978.
2. Sofronas, Anthony, Case Study No. 65, Taking risks and making high level presentations, *Hydrocarbon Processing*, November 2011; see also http://mechanicalengineeringhelp.com.
3. Sofronas, Anthony, *Analytical Troubleshooting of Process Machinery and Pressure Vessels Including Real-World Case Studies*, John Wiley & Sons, Hoboken, NJ, 2006.

Index

Note: Page numbers followed by *f* denote figures; page numbers followed by *t* denote tables.